Springer Theses

Recognizing Outstanding Ph.D. Research

Aims and Scope

The series "Springer Theses" brings together a selection of the very best Ph.D. theses from around the world and across the physical sciences. Nominated and endorsed by two recognized specialists, each published volume has been selected for its scientific excellence and the high impact of its contents for the pertinent field of research. For greater accessibility to non-specialists, the published versions include an extended introduction, as well as a foreword by the student's supervisor explaining the special relevance of the work for the field. As a whole, the series will provide a valuable resource both for newcomers to the research fields described, and for other scientists seeking detailed background information on special questions. Finally, it provides an accredited documentation of the valuable contributions made by today's younger generation of scientists.

Theses are accepted into the series by invited nomination only and must fulfill all of the following criteria

- They must be written in good English.
- The topic should fall within the confines of Chemistry, Physics, Earth Sciences, Engineering and related interdisciplinary fields such as Materials, Nanoscience, Chemical Engineering, Complex Systems and Biophysics.
- The work reported in the thesis must represent a significant scientific advance.
- If the thesis includes previously published material, permission to reproduce this must be gained from the respective copyright holder.
- They must have been examined and passed during the 12 months prior to nomination.
- Each thesis should include a foreword by the supervisor outlining the significance of its content.
- The theses should have a clearly defined structure including an introduction accessible to scientists not expert in that particular field.

More information about this series at http://www.springer.com/series/8790

Holly Alexandra Tetlow

Theoretical Modeling of Epitaxial Graphene Growth on the Ir(111) Surface

Doctoral Thesis accepted by
the King's College London, UK

Author
Dr. Holly Alexandra Tetlow
Department of Physics
King's College London
London
UK

Supervisor
Prof. Lev Kantorovich
Department of Physics
King's College London
London
UK

ISSN 2190-5053 ISSN 2190-5061 (electronic)
Springer Theses
ISBN 978-3-319-88140-9 ISBN 978-3-319-65972-5 (eBook)
DOI 10.1007/978-3-319-65972-5

Printed on acid-free paper

This Springer imprint is published by Springer Nature
The registered company is Springer International Publishing AG
The registered company address is: Gewerbestrasse 11, 6330 Cham, Switzerland

Supervisor's Foreword

Graphene, as a single layer of graphite, has been known for some time, mostly as a theoretical possibility. However, starting from 2004–2005, when it became clear that this material can actually be created in a laboratory [after scientists in Manchester (UK), Atlanta and New York (US) were able to obtain it], the real explosive growth of research began. This is because of graphene remarkable electronic, mechanical, transport and optical properties, which are due to a honeycomb lattice arrangement of carbon atoms.

First successful attempts to obtain this miracle material in a Manchester laboratory by A.K. Geim and K.S. Novoselov (for which they were awarded the 2004 Nobel Price in Physics) were based on a mechanical exfoliation (the scotch tape) method, which enabled them to produce only small quantities of graphene. Practical applications require, however, large amounts of this material, preferably with as little number of defects as possible. Epitaxial methods provide such an alternative. In epitaxial methods, hydrocarbon molecules when deposited on a metal surface decompose upon heating, hydrogen atoms evaporate and remaining carbon atoms nucleate and, under certain conditions, form hexagonal graphene flakes. Although considerable progress has been achieved in the direction of epitaxial methods in growing graphene, mostly experimentally, further advance would require understanding of detailed mechanisms at play, at atomistic level, of various stages of the whole process, starting from decomposition of hydrocarbon molecules, formation of carbon-only species (e.g. monomers or dimers) on the substrate, nucleation of carbon clusters and finally, their growth into hexagonal islands. Such a detailed microscopic understanding of the graphene growth by epitaxial methods, which can only be achieved by both theory and experiment, guiding each other, was still lacking at the time of the commencement of this thesis work.

The aim of this thesis was to consider, using theoretical methods rooted in first principles density functional theory [and in close collaboration with an experimental group of Prof. A. Baraldi (Trieste, Italy)], the following two initial stages of growth, considering ethylene molecules as the hydrocarbon species: (i) decomposition of ethylene on the surface of iridium, served as a substrate, which was shown to lead to carbon monomer gas on the surface, and, subsequently, (ii) nucleation of carbon monomers at high temperatures into clusters that serve as precursors of hexagonal graphene islands.

The decomposition problem required considering a rather large set of chemical reactions on iridium surface whereby ethylene molecules were gradually losing their hydrogen atoms with subsequent breakage of their carbon–carbon bond. This work required calculating transition rates for all the reactions and then performing kinetics simulations, in which all the species are allowed to diffuse, decompose, as well as compose again when are in proximity of each other; in addition, diffusing hydrogen atoms are considered forming hydrogen molecules on the surface and then evaporating from it. This work resulted in theoretical predictions for the time evolution of various hydrocarbon species on the surface during the epitaxial process, which were compared with XPS data obtained experimentally. Remarkably, in this work experiment and theory went side-by-side guiding each other, and without this close collaboration, no progress could have been made.

The main result of the decomposition simulations was that carbon monomers, rather than carbon dimers, appear in the end as the main species on the surface. These monomers form a two-dimensional gas in which atoms diffuse very quickly and hence collide forming carbon clusters. However, not all clusters are stable. Most of them would decompose the next moment after they are formed as adding more atoms to them makes them less energetically stable; only clusters of certain critical size will survive and continue growing. Kinetically formation of such clusters represents an extremely rare event. However, once such a cluster was formed, it serves as a nucleus for further growth as adding more atoms to it becomes energetically favourable. In this thesis, thermodynamic aspects of nucleation of hexagonal carbon clusters on the iridium surface at high temperatures were considered theoretically. We find that the free energy of cluster formation has a nontrivial form, quite different from the one assumed in classical nucleation theory. Moreover, our calculations were able to explain an experimental result that clusters of 5–6 carbon atoms play a pivotal role in the nucleation process on iridium after ethylene exposure. Some limited kinetic aspects of nucleation were also considered, opening a clear direction for further work.

I hope the work presented in this thesis will stimulate further research in the exciting area of graphene (and other two-dimensional materials) growth.

London, UK
July 2017

Prof. Lev Kantorovich

Abstract

The main problem affecting the widespread use of graphene based products is the ability to produce high-quality graphene in large quantities. One possible method of achieving this is to grow it epitaxially. In this thesis, a selection of the important processes involved in epitaxial graphene growth is studied in detail using density functional theory based calculations. It begins with an investigation into the initial stages of the growth process, specifically determining the kinetics of the decomposition of ethylene on the Ir(111) surface, in order to find the decomposition mechanism and the resulting carbon feedstock species that will go on to form graphene. To achieve this, the energy barriers of the relevant reactions are determined using the nudged elastic band (NEB) method, and then the reaction kinetics are modelled using both rate equations and a specifically developed kinetic Monte Carlo code. The decomposition is determined for a variety of different experimental conditions, including temperature programmed growth and chemical vapour deposition. Broadly, the results show that the decomposition mechanism involves the breaking of the C–C bond, resulting in the production of C monomers.

Following from this, the nucleation of carbon clusters on the Ir(111) surface from C monomers prior to graphene formation is investigated. The full, temperature dependent work of formation is devised and calculated for a variety of different cluster types. From this value, it is possible to determine the critical cluster size, where the addition and removal of C monomers is equally likely. Based on this, small arch-shaped clusters containing four to six C atoms are predicted to be long-lived on the surface, suggesting that they may be key in graphene formation.

Finally, the healing of single vacancy defects in graphene on the Ir(111) surface is examined. These defects are undesirable and negatively affect the useful properties of graphene. The attempted healing of the defects by ethylene molecules is simulated with molecular dynamics and used to predict partially healed structures. The energy barriers to the healing are determined using NEB calculations. The results suggest that the vacancy defects can be healed directly by dosing with ethylene molecules during graphene growth.

Parts of this thesis have been published in the following journal articles

- Growth of epitaxial graphene: Theory and experiment, H. Tetlow, J. Posthuma de Boer, I.J. Ford, D.D. Vvedensky, J. Coraux, L. Kantorovich, Physics Reports, 542 (2014) 195–295.
- Ethylene decomposition on Ir(111): Initial path to graphene formation, Holly Tetlow, Joel Posthuma de Boer, Ian J. Ford, Dimitri D. Vvedensky, Davide Curcio, Luca Omiciuolo, Silvano Lizzit, Alessandro Baraldi, and Lev Kantorovich, Physical Chemistry Chemical Physics 18 (2016) 27897–27909.
- A free energy study of carbon clusters on Ir(111): Precursors to graphene growth, H. Tetlow, I. J. Ford, and L. Kantorovich, Journal of Chemical Physics 146 (2017) 044702.
- Hydrocarbon decomposition kinetics from first principles H. Tetlow, L. Kantorovich, In Progress.

Acknowledgements

First, I would like to thank my supervisor Prof. Lev Kantorovich for all his help and guidance in the last 4 years. I have been lucky to have such a good supervisor and without his help and patience all this might have been impossible.

I would like to acknowledge my collaborators from the Thomas Young Centre in London for their help and advice with the hydrocarbon decomposition work. These include Joel Posthuma de Boer (Imperial College London), Ian J. Ford (University College London) and Dimitri Vvedensky (Imperial College London). Joel also produced the code for the rate equations which is included in this thesis. In addition, the people listed above, along with Johann Coraux (CNRS, Institut NEEL, Grenoble), were responsible for the review we produced for Physics Reports titled "Growth of epitaxial graphene: Theory and experiment".

The experimental work on the thermal decomposition of ethylene was completed by Prof. Alessandro Baraldi's group at the University of Trieste and Elettra—Sincrotrone Trieste. I would like to thank all the group members involved in this: Davide Curcio, Luca Omiciuolo and Silvano Lizzit for their help and patience. These experiments were extremely valuable for the majority of the work in this thesis.

The HPC simulations in this would not have been possible without the use of the HECTOR and ARCHER supercomputers and all those who maintain the facility. I would like to thank the EPRSC and the material chemistry consortium (MCC) for providing access to these supercomputers.

Finally, I would like to acknowledge my funding source, the Engineering and Physical Sciences Research Council (EPRSC).

I would like to thank my friends and family who have supported me over the last 4 years. Especially my parents who have done their best to keep me out of trouble and got me through the hard times. I am grateful that I am still in good touch with the friends that I met during my Undergraduate degree at Bristol and that we have continued to bond over sharing the "Ph.D. experience". I would also like to thank my new friends that I have met in London (esp. Felicia) for making living here much more fun. This includes my awesome boyfriend Paul whose constant love and support has been a source of constant love and support.

Finally, I would like to thank my 5-year-old laptop. It likes to overheat a lot and really enjoys taking intermittent "breaks" but it has survived my Ph.D. and the writing of this thesis. I am very grateful that that its hard drive has only decided to fail once, even though the timing was awful.

Contents

Chapter 1
Review of Epitaxial Graphene Growth

1.1 Epitaxial Graphene Growth

Graphene is the so-called "miracle material" of the 21st century [1]. Since the Nobel prize was awarded to Novoselov and Geim in 2010 for their graphene research it has attracted large amounts of interest, both in the scientific community and in the public [2, 3]. This is brought about due to its unique electronic, mechanical and optical properties. Its thermal conductivity, tensile strength and charge carrier mobility are all significant when compared to current materials [4–6]. In the lab these properties were discovered and tested by producing graphene by the exfoliation of graphite [2, 3]. Typically scotch tape was used to repeatedly remove layers from the graphite surface to leave a single graphene layer. This method, although suitable for creating small, high quality graphene samples is not suitable for producing graphene on a large scale. Hence one of the biggest challenges preventing the widespread development of graphene technology is being able to produce it in large quantities with adequate quality.

A promising way of producing graphene on a more general scale is by growing it epitaxially. Epitaxial graphene growth usually involves depositing a carbon source onto the surface of a transition metal and then using high temperatures (up to 1200 K) to facilitate the formation of graphene islands. However how graphene forms depends heavily on the growth conditions. These factors affect the structure of the graphene produced and can result in undesirable defects and grain boundaries. They may also, however, be used to tailor the growth process, so that high quality graphene is formed. To do this the underlying mechanisms involved on an atomic scale must be fully understood.

The epitaxial graphene growth process begins when a source of carbon is deposited onto the transition metal surface. This may be at room temperature, prior to heating as in temperature programmed growth (TPG), or with a constant flux at high temperatures as in chemical vapour deposition (CVD) [7]. In temperature programmed growth the temperature is then increased to facilitate the growth. In both cases if a

H.A. Tetlow, *Theoretical Modeling of Epitaxial Graphene Growth on the Ir(111) Surface*, Springer Theses, DOI 10.1007/978-3-319-65972-5_1

hydrocarbon precursor is used it must decompose on the metal substrate to form some kind of carbon feedstock, which will then act as the building blocks for graphene growth.

The next step towards growth is the nucleation of carbon clusters. Once the products of the decomposition are produced they can diffuse freely on the surface. Under the influence of high temperatures nucleation will occur via the combining of these small carbon species to form larger and larger clusters. This process may be assisted by the substrate structure, whereby defects and step edges can trap atoms and molecules, and increase the likelihood of nucleation. The carbon clusters that nucleate may then be able to diffuse and coalesce to eventually form graphene islands.

Finally, a layer of graphene is produced on top of the substrate. At this stage the graphene will likely contain undesirable defects which reduce the conductivity. However it may be possible to heal vacancy defects in graphene by dosing the surface with hydrocarbons in order to increase the quality [8].

This thesis will uncover the atomistic processes involved in epitaxial graphene growth, focusing on the initial stages of growth, where hydrocarbons decompose to form a carbon feedstock, and the nucleation of this feedstock on the substrate, leading to the formation of carbon clusters. It will then discuss healing of defective graphene once it has been grown, made possible by dosing the surface with hydrocarbons. The formation of graphene islands, and their subsequent growth by the addition of carbon species will not be considered.

The text will begin with a review of previous studies of epitaxial graphene growth, which will discuss the current experimental and theoretical understanding of the growth process. Then the relevant calculation methods and techniques will be described. The main areas of focus, outlined above will be presented in detail along with the relevant results and conclusions. It will conclude with a summary of the main results, and an outlook for the future of graphene growth.

1.1.1 Experimental Techniques

In order to study the graphene growth process a variety of experimental techniques are employed. Some of these allow atomic resolution of the surface. Such techniques may allow *in operando* analysis whereby the system may be studied throughout the growth process, requiring the use of a ultra high vacuum (UHV). Alternatively with *ex situ* analysis the system is characterised outside of the growth environment. This may require cooling the system to room temperature before any measurement can be taken. *In operando* analysis is more suitable for imaging graphene growth since it allows the growth process to be resolved. Table 1.1 below summarises the main experimental methods, which are discussed within this thesis. A brief summary of each method is also included below [9].

Table 1.1 Experimental techniques used for characterising graphene growth and their operating conditions. Adapted from [9]

Name	Type	Spatial resolution	Operating temperature [K]	*In operando*?
Scanning tunnelling microscopy (STM)	Scanning probe	≤Å	<1–1000	Yes
Atomic force microscopy (AFM)	Scanning probe	≤Å in UHV	4–300	No
Low energy electron diffraction (LEED)	Electron scattering	≈100 μm	300–1500	Yes
Reflection high energy electron diffraction (RHEED)	Ensemble average based on electron scattering	≈100 μm	300–1500	Yes
Ramen spectroscopy	Ensemble average	≈100 nm	4–700	No
Scanning electron microscopy (SEM)	Scanning probe	≈nm	300	No
X-ray photoemission spectroscopy (XPS)	Ensemble average based on electron excitation	≈100 μm	4–1100	Yes

1.1.1.1 Scanning Tunnelling Microscopy (STM)

In scanning tunnelling microscopy a sharp tip is used to scan the surface of a conducting material. Electrons can tunnel from/to the tip to/from the surface. The tunnelling current decays exponentially as a function of the tip-sample distance and is extremely sensitive to the lateral position of the tip. This can be used to achieve atomic resolution. Two imaging methods are used: constant current or constant height. In constant current mode the current is fixed and the height of the tip with respect to the sample is allowed to vary, via the use of a constant feedback loop which ensures the current remains fixed. In this way the change in the tip-sample distance can be monitored. In constant height mode the height is fixed and the current varies across the sample.

1.1.1.2 Atomic Force Microscopy (AFM)

In atomic force microscopy a scanning tip attached to a cantilever is used to scan the sample. The tip detects variation in the short-range atomic forces across the sample corresponding to the surface topology. The forces are transformed into an electric signal corresponding to the movement of the cantilever. The total force consists of many different contributions which can be attractive or repulsive such as, van der Waals, electrostatic, ionic, frictional, chemical or capillary forces. AFM operates in either static or dynamic mode. In static mode the tip is deflected by weak forces, requiring a soft cantilever. This mode requires that the eigenfrequencies of the forces are suitably different from the resonance frequency of the cantilever. In dynamic mode the tip is made to oscillate with either constant frequency or constant amplitude (or both) with the resonance frequency maintained. In the former case the change in the amplitude and phase of the tip is measured. In the latter case either the tip height can be measured while the resonance frequency is kept constant, or the height can be kept constant while the change in the resonance frequency is monitored. This is known as non-contact mode (NC-AFM).

1.1.1.3 Low-Energy Electron Diffraction (LEED) and Reflection High-Energy Electron Diffraction (RHEED)

LEED and RHEED are both used to study highly crystalline surfaces. LEED involves firing low energy (<100 eV) electrons at the sample and then observing the scattering events. This allows the reciprocal lattice of the upper layer to be imaged. Electrons in LEED are usually back-scattered. In RHEED the electrons are high energy (≈ 10 keV) and can penetrate deeper into the sample. Once scattered the electrons form a diffraction pattern according to the features of the surface. In RHEED the electrons are more commonly forward scattered than in LEED. RHEED is also ideal for *in operando* measurements and allows access to the sample during the diffraction measurements.

1.1.1.4 Raman Spectroscopy

Raman spectroscopy measures the inelastic scattering of electromagnetic waves based on the Raman effect. When applied to a sample the electromagnetic waves can gain or lose energy by the excitation of phonons in the material. This energy change can be used to characterise the atomic elements and their chemical bonding (e.g. interatomic or van der Waals) via proper theoretical modelling. The scattered electromagnetic waves are filtered to remove a component due to the elastic scattering and the rest is fed into a detector.

1.1.1.5 Scanning Electron Microscopy (SEM)

The scanning electron microscope scans a sample with a focused beam of electrons. These typically have a energy between 100 eV to a few keV. The electrons produce secondary electrons which are emitted from excited atoms close to the surface of the sample. These allow imaging of the surface topography. The interaction of the electron beam with the sample also causes some of the electrons to be back-scattered. These scatter from the sample elastically with high energy and often penetrate deep into the sample.

1.1.1.6 X-Ray Photoemission Spectroscopy (XPS)

X-ray photoemission spectroscopy uses the photoelectron effect in order to probe the surface of a sample. X-ray photons are applied to the sample where their energy can excite core level electrons associated with atoms on the surface. This will occur if the photon energy is greater than that of the sample work function. The kinetic energy of the excited electrons is then measured. The binding energy of the core electron can be found from the energy difference between the kinetic energy of the excited electron and the work function. The core level electron binding energy depends on the atomic element of the atom it originated from and also on its chemical bonding (the local environment). Therefore it can be used to identify the type of species on the surface as well as their environment.

1.2 The Graphene Growth Process

1.2.1 Producing a Carbon Source

Typically in epitaxial graphene growth hydrocarbon molecules can be used as a source of carbon. They are deposited onto the metal surface where the catalytic action of the surface and high temperatures allow them to decompose to form carbon based building blocks, which then go on to form graphene. In CVD growth the system is initially at high temperatures and the molecules arrive with some kinetic energy, whereas in TPG growth the molecules are deposited at room temperature before the temperature is increased. In each case the decomposition of the hydrocarbon molecules is the first stage towards graphene growth.

It is obvious that during the decomposition process hydrocarbon molecules must dehydrogenate to leave the carbon species that are required for forming graphene. However the form that these carbon species take after this is not obvious. For instance a hydrocarbon may completely decompose into carbon monomers or partially decompose so that only hydrogen atoms are removed. This will be dependent on the type of hydrocarbon supplied as well as the substrate and the experimental conditions. For graphene growth many different hydrocarbons are used, such as methane, ethylene or benzene. The choice of precursor will effect the temperature range when graphene is formed based on the energy required to break the necessary bonds.

This was demonstrated experimentally in [10]. Graphene was grown on a Cu foil using various precursors. The Raman spectroscopy (a) and SEM images (b,c) are shown in Fig. 1.1. It was shown that when benzene is used graphene is first identified at 573 K, compared with 1073 K for methane. This is was suggested to be because less energy is required to dehydrogenate benzene than methane, hence the carbon species that form graphene can be produced at lower temperatures. Also the structure of benzene contains a hexagonal carbon ring, whereas methane has only one C atom. If the benzene molecules do not completely break down into C monomers then larger

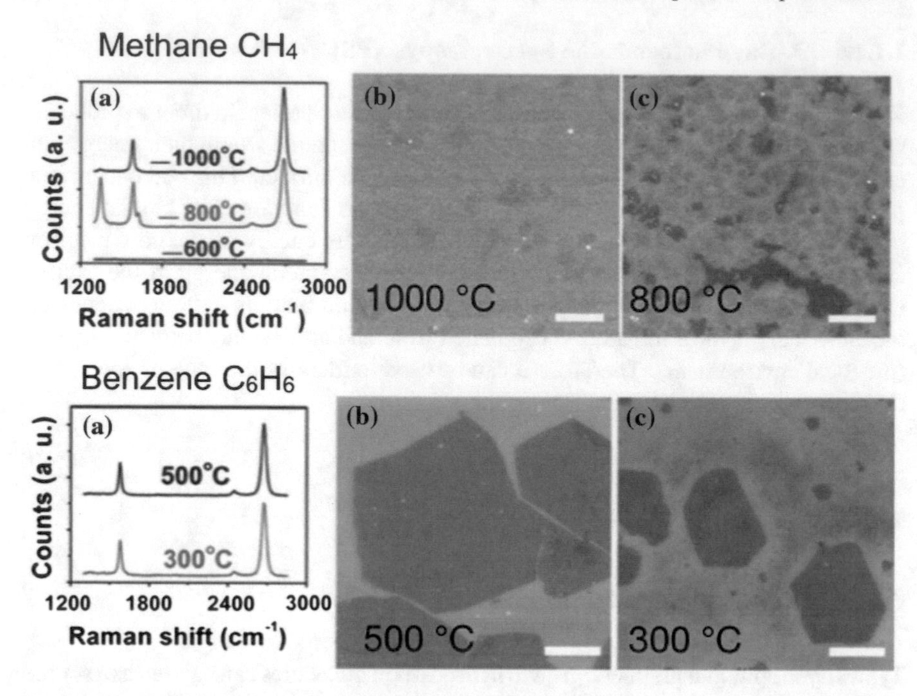

Fig. 1.1 The Raman spectroscopy and SEM images for graphene growth with methane and benzene. When methane is used as a precursor graphene islands form above 800 C (1073 K), whereas when using benzene islands are present at 300 C (573 K). Reprinted (adapted) with permission from [10]. Copyright (2011) American Chemical Society

carbon species can remain, which can produce graphene at a more rapid rate. Methane on the other hand will dehydrogenate to produce C monomers.

By using in-situ x-ray photoemission spectroscopy (XPS) experiments it is possible to monitor the decomposition of hydrocarbons as the temperature increased. The core level shift from the electrons in C atoms in the hydrocarbon species gives rise to peaks in the spectra, whose energies depend on the environment of the C atom. Identifying particular species from the peaks is possible by using multiple techniques. Firstly each different environment of a C atom in a particular molecule will give rise to an XPS peak, hence the number of peaks appearing at the same time should roughly correlate with the number of C environments. However phonon excitations may produce vibrational satellites

For methane deposition on the Pt(111) surface the thermal evolution of species was found in this way [11]. Using CVD methane was deposited onto the surface, and then the temperature of the system was ramped up linearly. The relative coverages of species on the surface is shown in Fig. 1.2a. Methane itself does not adsorb onto the surface. Instead it can immediately dehydrogenate to CH_3 when arriving at the surface due to its kinetic energy. Above 250 K CH_3 dehydrogenates to CH. Simultaneously some of the CH_3 is lost from the surface via hydrogenation to methane. Once the

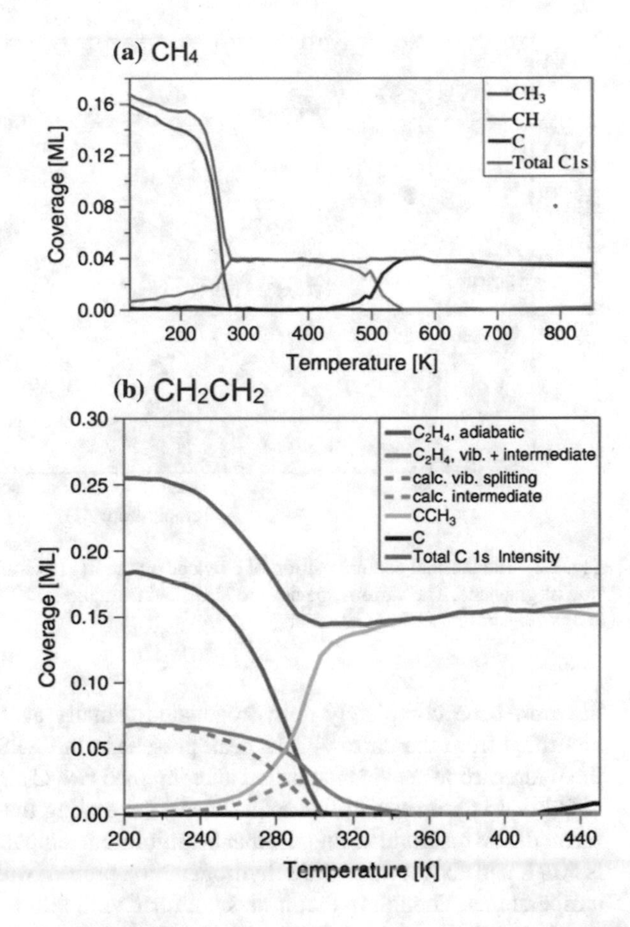

Fig. 1.2 The thermal evolution of hydrocarbon species on Pt(111) starting from the deposition of **a** methane and **b** ethylene. The coverages are determined from in-situ XPS spectroscopy [11]. Copyright (2011) American Chemical Society

temperature reaches 450 K CH dehydrogenates to form carbidic species, C_n which contain no hydrogen.

For ethylene the situation is more complicated as many different reactions are possible. Hydrogen may be removed from or added to either of the C atoms. Furthermore there is the possibility that the C-C bond may break. The decomposition of ethylene on Pt(111) was also studied in [11]. After depositing ethylene onto Pt(111) at a pressure of 1.7×10^{-8} mbar, it was reported that CH_3C (ethylidene) was produced after heating to 250–280 K. An intermediate species was also identified in this temperature range (Fig. 1.2b blue dashed line). This is formed during the conversion of CH_2CH_2 to CH_3C and is suggested to be CH_3CH due its XPS signal. At 420 K carbidic species are found.

For ethylene deposited on Ir(111) the entire decomposition path prior to graphene formation was determined using XPS [12]. The evolution of the surface coverage is shown in Fig. 1.3. The reaction process begins with ethylene transforming to CH_3CH. Then three H atoms are sequentially removed to leave CH_3C and then CHC, which exists in two forms. Above 500 K carbidic carbon is found, suggesting that ethylene

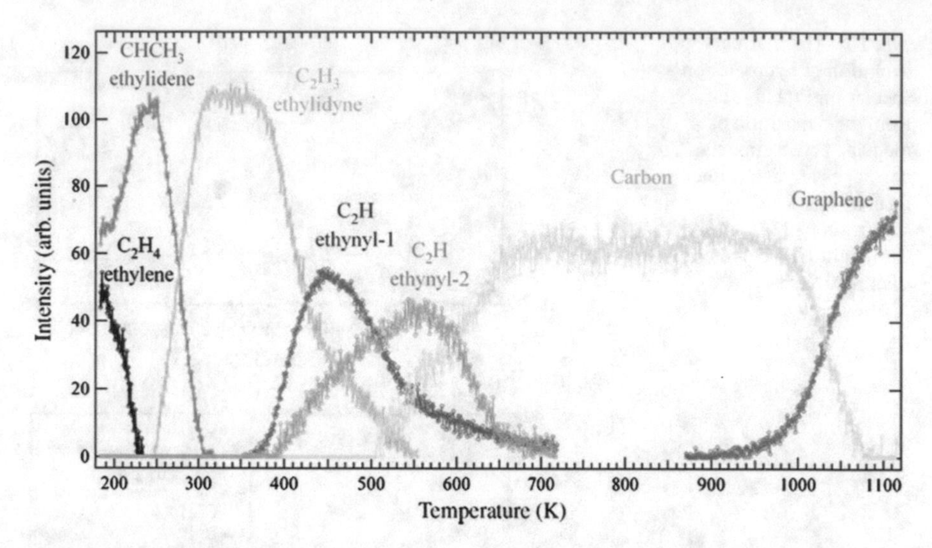

Fig. 1.3 The thermal decomposition of ethylene on the Ir(111) surface leading towards the formation of graphene. The various species are identified from in-situ XPS experiments. Reprinted from [12] with permission from Elsevier

has now been completely dehydrogenated. Finally at 950 K graphene is formed, identified from the narrow C 1s peak present in the XPS spectra. It is of note that the hydrocarbon species identified all contained two C atoms, hence breaking of the C-C bond is not seen in this experiment, suggesting that C monomers will not be formed. The results also suggest that the minimum temperature for forming graphene is 500 K and above, since the dehydrogenation process was not complete below these temperatures. Though in fact it dies not form until 900 K.

The XPS experimental results presented above are successful in imaging the change in concentration of the various species on a substrate as the temperature is increased. However a disadvantage of these methods is that they are not able to resolve the entire process, especially if reactions happen quickly so that some species can only exist on very short timescales. Also the assignment of peaks to particular species may be difficult due to the fact that there are many possible reactions, and hence many possible species need to be considered. By using theoretical density functional theory (DFT) based methods it is possible to consider multiple competing processes and then predict the mechanism for the decomposition based on the energetics of the relevant reactions. Typically in these methods the energy barriers for the various reactions are calculated (often using the nudged elastic band method, which is explained in detail in Sect. 2.4). These can then be used to find which reactions are most likely, and even the kinetics can be calculated to predict the sequence of reactions. In order to determine the correct sequence it is important that all possible reactions starting from the initial hydrocarbon molecule are considered.

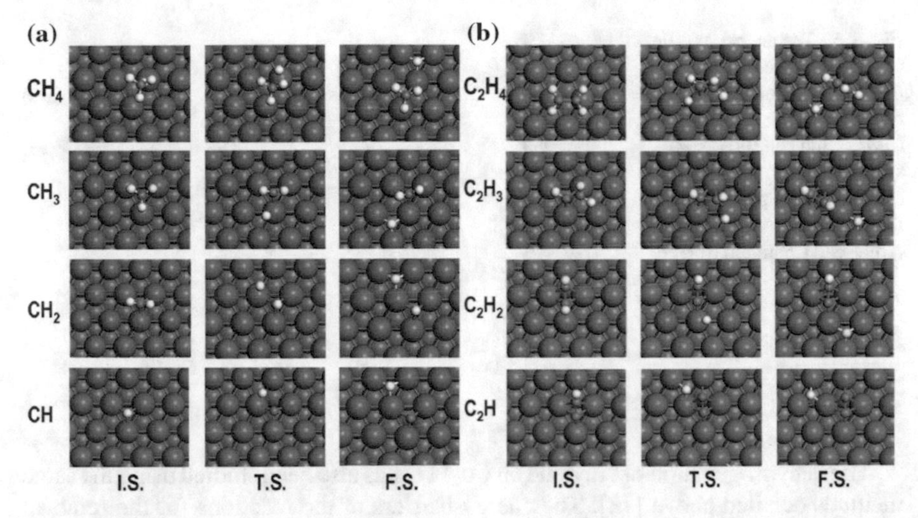

Fig. 1.4 The structures of the initial (I.S.), transition (T.S.) and final (F.S.) states corresponding to the direct processes leading to complete dehydrogenation of **a** methane and **b** ethylene on the Cu(111) surface. The *red, white* and *grey spheres* represent Cu, H and C atoms, respectively. Reprinted (adapted) with permission from [13]. Copyright (2011) American Chemical Society

The simplest hydrocarbon to consider in this way is methane. In this case once the molecule has adsorbed onto the substrate the basic method of the decomposition must involve the sequential removal of H atoms to leave a carbon monomer with the reaction sequence: $CH_4 \rightarrow CH_3 + H \rightarrow CH_2 + 2H \rightarrow CH + 3H \rightarrow C + 4H$.

In the work of Zhang et al. the energy barriers for removing each H atom from methane on the Cu(001) and Cu(111) surfaces were calculated [13]. For each reaction DFT calculations were used to find the minimum energy configurations for the initial (reactant) and final (product) states. The initial state (I.S.) consists of the complete CH_x molecule and the final (F.S.) state consists of the molecule with an H atom removed onto the surface (CH_{x-1} + H). These are shown in Fig. 1.4a. The DFT-based nudged elastic band method was used to find the transition state (T.S.) which corresponds to the saddle point on the minimum energy pathway connecting the two states. By finding the energy difference between the minima and the saddle point the energy barriers are found. The energy profile for all the methane dehydrogenation reactions on the Cu(001) and Cu(111) surfaces is shown in Fig. 1.5.

The results indicate that dehydrogenation of methane is highly unfavourable on both surfaces. The energy barrier for removing an H atom is 1–2 eV at each dehydrogenation step and always results in the final state having a greater energy than the initial state. This effect is such that the final product C+4H is either 2.75 eV (Cu(111)) or 3.60 eV (Cu(001)) greater in energy than the initial methane molecule. These large barriers show that if methane is used as a graphene growth source high temperatures will be required in order to facilitate its decomposition. However, other authors have determined that if two C monomers go on to form C_2 after the dehydrogenation process then the overall energy change is lower with respect to methane [14].

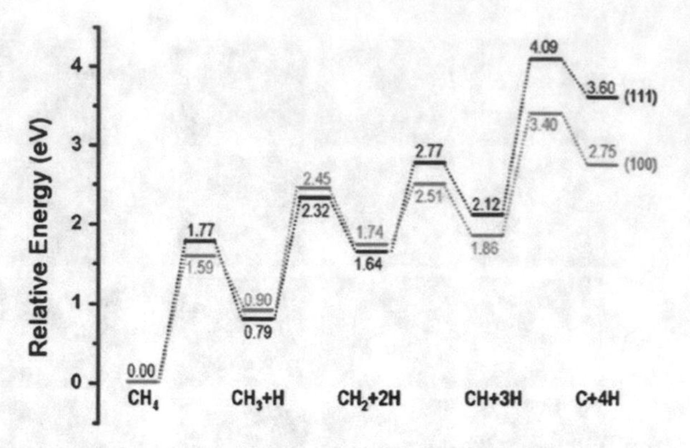

Fig. 1.5 The energy profile of the complete direct dehydrogenation process of methane on the Cu(111) (*black*) and Cu(100) (*red*) surfaces. Reprinted (adapted) with permission from [13]. Copyright (2011) American Chemical Society

The dehydrogenation of ethylene on Cu(111) has also been studied using the same methods detailed above [13]. The energy barriers of the reactions for the removal of four H atoms from the molecule to leave a carbon dimer were calculated. The reaction sequence was $CH_2CH_2 \rightarrow CH_2CH + H \rightarrow CHCH + 2H \rightarrow CHC + 3H \rightarrow CC + 4H$ (Fig. 1.4b). As with the removal of H from methane all the energy barriers were large (between 1–2 eV), leading to the conclusion that forming a C dimer from ethylene is unfavourable. However this is a very simplistic treatment of ethylene decomposition. In actual fact, since there are two C atoms in ethylene H atoms can be removed from or added to either carbon atom. Furthermore there is the possibility to break the C-C bond of any of the C_2H_m species. As a result many more species and reactions are possible. This will result in many different mechanisms for producing carbon.

More complicated ethylene dehydrogenation mechanisms have been considered using similar methods. For ethylene adsorbed on the Pt(111), Pd(111) and Rh(111) surfaces the removal of one H atom to form ethylidyne (CH_3C) was investigated with kinetic Monte Carlo (kMC) simulations, including the possibility for both hydrogenation and dehydrogenation reactions [15–17]. As a result alternative mechanisms are possible. In this manner the authors produced a reaction scheme, shown in Fig. 1.6. Here arrows pointing to the left represent dehydrogenation reactions, whereas arrows pointing to the right represent hydrogenation reactions. Solid downward and upward arrows show adsorption and desorption reactions respectively and the dashed downward arrows represent the possible isomerisation reactions.

The energy barriers for each reaction (excluding isomerisation reactions) were calculated with DFT methods. The barriers are then used to determine the reaction rates. By using kinetic Monte Carlo (kMC) simulations the preferred reaction mechanism for the conversion of ethylene to CH_3C (ethylidyne) was found on each surface. The three competing mechanisms shown in Fig. 1.6, are: 1. $CH_2CH_2 \rightarrow CH_2CH \rightarrow CH_3CH \rightarrow CH_3C$, 2. $CH_2CH_2 \rightarrow CH_2CH \rightarrow CH_2C \rightarrow CH_3C$, and 3. $CH_2CH_2 \rightarrow CH_3CH_2 \rightarrow CH_3CH \rightarrow CH_3C$. As well as dehydrogenation and hydrogenation reactions the desorption of hydrogen to form hydrogen gas is included in the kMC simulation. This is an important consideration since the amount of H atoms on the

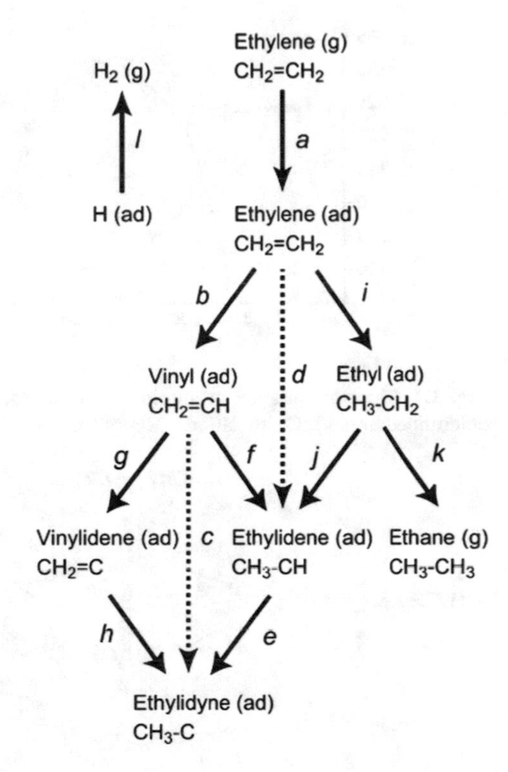

Fig. 1.6 The possible mechanisms for the removal of a hydrogen atom from ethylene on Pt(111) and Pd(111). The *left* and *right downward arrows* represent dehydrogenation and hydrogenation reactions respectively whereas the *dashed downward arrows* represent the possible isomerisation reactions. For ethylene and hydrogen the *solid downward* and *upward arrows show* adsorption and desorption reactions respectively. Reprinted from [17] with permission from Elsevier

surface affects the other processes. A low availability of H atoms on the surface will limit hydrogenation reactions and will alter the reaction pathway.

For both Pt(111) and Pd(111) the preferred mechanism to ethylidyne was via CH_2CH and CH_2C (mechanism 2). For Pt(111) the rate limiting step in this process is the final hydrogenation reaction $CH_2C + H \rightarrow CH_3C$. This is due to the low availability of H atoms on the surface since they are easily lost from the formation of H_2 molecules. For Pd(111) the dehydrogenation of ethylene is the limiting process due to its high energy barrier. Once this reaction is completed CH_3C is almost instantly formed since the following reactions can happen quickly. This is because there energy barriers are comparatively lower and H atoms are readily available. The change in concentration for the various species on the surface as the temperature is increased as determined from the kMC simulations is shown in Fig. 1.7 for (a) Pd(111) and (b) Pt(111).

The work detailed above shows that the kinetics of many competing reactions can be simulated in order to find the most probable reaction pathway. However the reaction scheme presented in these papers (Fig. 1.6) is incomplete. Firstly only the removal of one H atom from ethylene was considered, and as a result many species with two H atoms or less are not included. Furthermore only hydrogenation and dehydrogenation reactions are possible, in fact isomerisation and C-C breaking reactions should be included since all possible reactions need to be accounted for in order to get realistic simulation results.

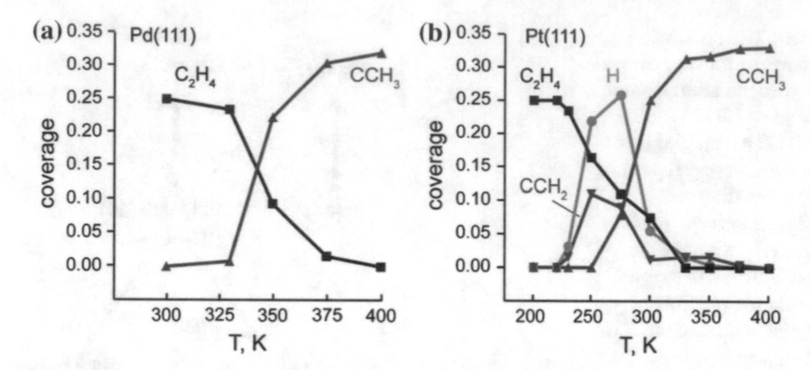

Fig. 1.7 The concentration of species on **a** the Pd(111) surface and **b** the Pd(111) surface as determined from kMC simulations. Reprinted from [17] with permission from Elsevier

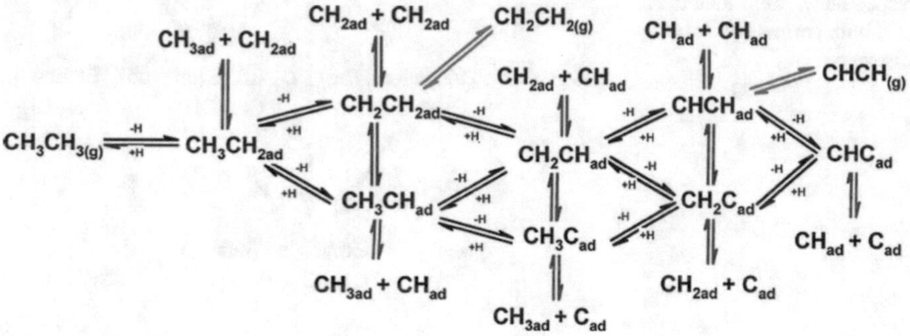

Fig. 1.8 The complete reaction scheme for C_nH_m species showing all possible reactions. These are hydrogenation/dehydrogenation reactions (*black arrows*), isomerisation reactions (*red arrows*), C-C breaking/recombination reactions (*blue arrows*) and adsorption/desorption reactions (*green arrows*). Reprinted (adapted) with permission from [18]. Copyright (2010) American Chemical Society

A complete reaction scheme for the decomposition of ethylene on Pt(111) and Pt(211) was presented in [18]. The diagram for this is shown in Fig. 1.8. Here all possible reactions and species have been considered, ranging from ethane (CH_3CH_3) to single C monomers (C_nH_m with $n = 1 - 2$ and $m = 0 - 6$).

The energy barriers for all of these reactions were calculated and used to predict the likelihood of various reactions. Based on these barriers it was concluded that species on the stepped Pt(210) surface are generally more stable than on the Pt(111) terrace. For both surfaces CH_3C is the most stable species. Isomerisation reactions were found to be unfavourable, whereas C-C breaking reactions were predicted to be feasible for some species, namely CH_3CH, CH_3CH_2 and CHCH.

These conclusions were based solely on the values of the energy barriers, with no calculation of the kinetics. When trying to determine the reaction sequence it is difficult to judge which reactions may occur based just on the barriers since we cannot predict how the concentration of the various species may change over time,

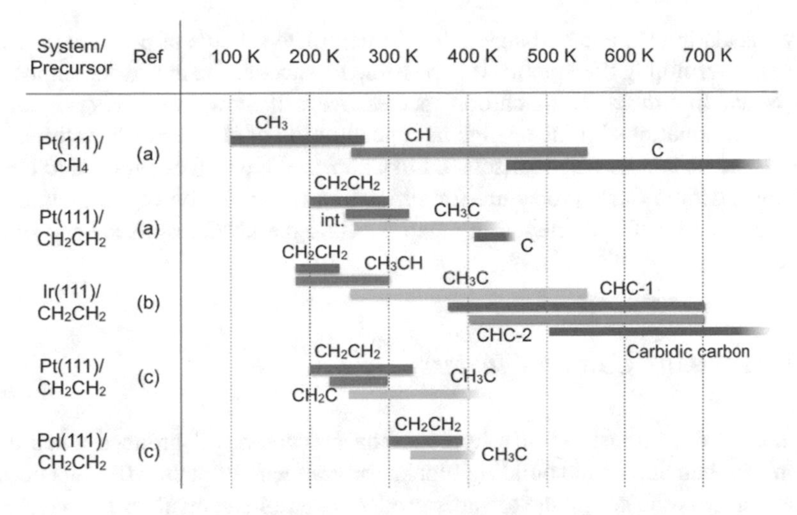

Fig. 1.9 The thermal evolution for different substrate/precursor systems as determined by experimental and theoretical studies. The references given correspond to, **a** [11] **b** [12] and **c** [17]

especially with temperature increases. This, including the concentration of H atoms will greatly affect which reaction pathways are accessible. Hence it is necessary to simulate the reaction kinetics in order to determine the decomposition reaction sequence.

A summary of the results from the experimental and theoretical studies of hydrocarbon decomposition reactions is presented in Fig. 1.9. The thermal evolution of the various species on the different substrates is shown.

The results show that CH_3C is often formed when ethylene is deposited. This is reported on the Pt(111), Pd(111) and Ir(111) surfaces for both experimental and theoretical studies and has also been widely recognised in other studies [11, 12, 17]. There is good agreement between the experimental and theoretical results on Pt(111) [11, 17]. Both determine that ethylene converts to CH_3C within a similar temperature window. The theoretical results find CH_2C as an intermediate species in this transition, whereas the experimental results recognise that there is an intermediate species, but interpret it as being CH_3CH.

After the formation of CH_3C in these studies there is limited knowledge about the mechanism for forming carbon. The kinetic Monte Carlo simulation results do not go beyond forming CH_3C, since the allowed reactions only include C_2H_m species containing two or more H atoms, and C-C breaking reactions are not possible. The experimental results indicate that carbidic species are eventually formed but the processes that occur between the formation of CH_3C and this stage are only vaguely determined. It is unknown whether the C-C bond is broken at some stage (as suggested in [18]) to allow the C monomers to be produced, or whether dimers are the prevalent species which go on to form carbidic carbon.

By calculating the energy barriers for the complete scheme of possible reactions and then determining the kinetics it is possible to uncover the entire decomposition process, and find the resulting carbon species. We will show that this is possible by solving rate equations for the change in concentration of all species over time. This method, and its results will be discussed in Chap. 3. We will then proceed to build a full kinetic Monte Carlo procedure for simulating the evolution of the system on a grid. The code itself is discussed in Chap. 4, while the kMC results are the subject of Chap. 5.

1.2.2 Forming Carbon Clusters

The thermal decomposition of a hydrocarbon precursor will produce a source of carbon that then acts as the building blocks for graphene growth. After this decomposition process the next stage towards producing graphene involves the nucleation of carbon clusters on the surface. These clusters will then grow in size until graphene islands are formed. This phase, characterised by the nucleation and growth of carbon clusters is important to the overall graphene growth process. Where the nucleation occurs will affect the overall quality of the graphene since too many nucleation sites gives rise to grain boundary defects when multiple growth fronts meet. Furthermore the stability of different clusters that are formed dictates how long they are likely to exist on the surface and hence whether they will be incorporated into the graphene growth front. Therefore in order to develop a clear understanding of the initial stages of graphene growth it is necessary to investigate the formation of carbon clusters on transition metal surfaces, starting from their nucleation.

In graphene growth experiments carbon species have been identified prior to the formation of graphene [7]. In some cases, with scanning tunnelling microscopy (STM) imaging the nature of these species can be identified. For example individual carbon clusters have been observed on the Ru(0001) and Rh(111) surfaces, before graphene starts to form [19, 20]. These carbon clusters were found to contain 13 C atoms or more arranged in multiple hexagonal rings, which form a dome-like shape on the surface. Images of these clusters are shown in Fig. 1.10.

These dome-like clusters have also been identified on the Ir(111) surface by using XPS experiments and DFT calculations [21]. These clusters interact strongly with the surface by their edge atoms, which causes their dome-like shape. It was determined in this work that the formation of these clusters is an intermediate step between carbidic carbon and the emergence of graphene. This is in agreement with observations in other growth experiments.

Before these dome-like carbon clusters can be formed, carbon monomers must nucleate on the surface and form small clusters, which then grow to become stable once they reach a particular size. The stability of small carbon clusters with different structures with up to 13 carbon atoms on various surfaces has been determined from DFT calculations. Typically in these studies the formation energy (or variant of) for each cluster is calculated (at zero temperature), and used to determine the stability of a cluster.

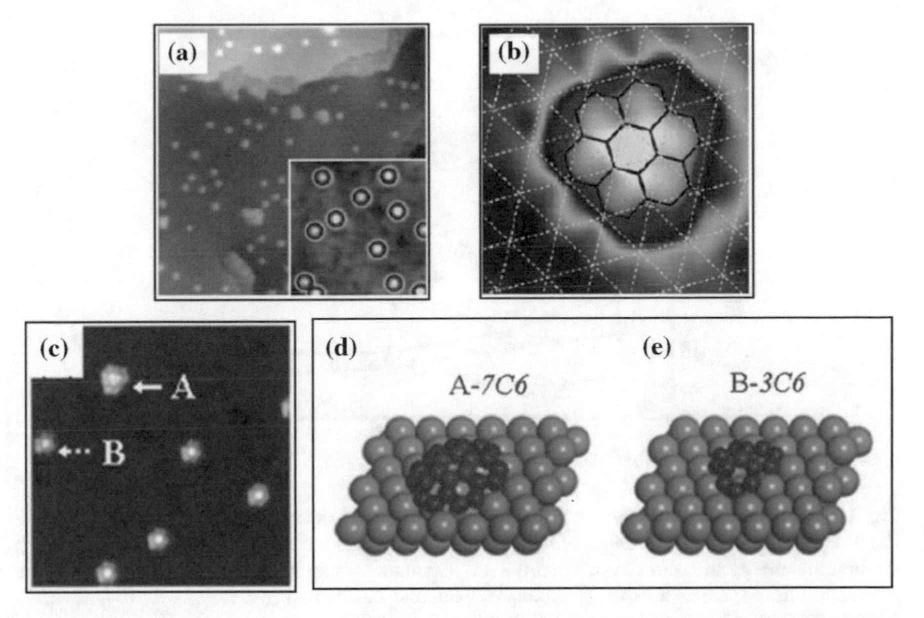

Fig. 1.10 Experimental evidence for carbon clusters on **a**, **b** Ru(0001) [19] and **c**, **d**, **e** Rh(111) [20]. For Ru(0001) a 24 C atom cluster as shown in **a** and **b** is found, whereas for Rh(111) both a 24 C atom cluster **d** and a smaller 13 C atom cluster **e** are observed as labelled in **c** as A and B respectively. **a**, **b** Reprinted (adapted) with permission from [19]. Copyright (2011) American Chemical Society. **c**, **d**, **e** Reprinted from [20] with permission from The Royal Society of Chemistry

Carbon monomers and dimers must exist prior to the formations of larger clusters. When nucleation occurs monomers become trapped on the substrate, potentially at step edges or in defects. Once trapped the monomers are immobile and then it will be easier for the growth to initiate. The adsorption (or formation) energies of dimers and monomers on the Cu(111), Ir(111) and Ru(0001) surfaces were calculated, for molecules on both terraces and step edges [22]. By calculating the binding energy of a dimer from two monomers on the terrace and at a step edge for each substrate it was determined where the preferable site for nucleation is. The binding energy for forming C dimers on the Ir(111), Ru(0001) and Cu(111) surfaces is shown in Fig. 1.11.

For Ru(0001) and Ir(111) dimers are not energetically stable on the terrace, and will decay to two monomers. However at the step edges dimers are more stable, and can become trapped. For Cu(111) this situation is reversed and dimers are stable on terraces. These results suggest that nucleation will begin at step edges for Ru(0001) and Ir(111), and at terraces for Cu(111). However only the stability of monomers and dimers was considered, which only reflects the very beginning of nucleation. For larger clusters the trend may differ, and even on terraces at some cluster size stability will be achieved. The role of temperature is also important since the mobility of the dimers and monomers will depend on it.

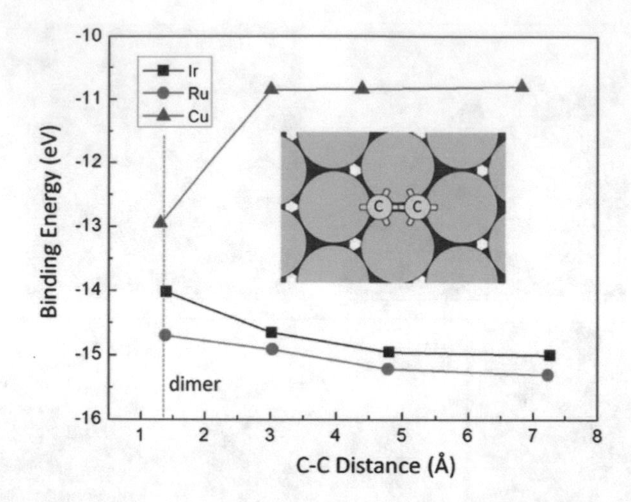

Fig. 1.11 The energetics of forming a C dimer from two monomers on the Ir(111), Ru(0001) and Cu(111) surfaces. On Ir(111) and Ru(0001) dimer formation is unfavourable, however on Cu(111) the binding energy of a dimer is around 0.2 eV less than two monomers. Reprinted figure with permission from [[22] as follows: H. Chen, W. Zhu, and Z. Zhang, Phys. Rev. Lett., 104, 186101, 2010.] Copyright (2010) by the American Physical Society

Beyond monomers and dimers the most stable clusters are often found to have specific structures, and can therefore be categorised into different types. For example linear clusters are typically most stable when forming arches, where the end atoms strongly interact with the surface and the inner atoms are raised up away from it. Compact clusters often have carbon atoms arranged in hexagonal rings to form sp^2 networks.

For clusters on the Cu(111) terrace the formation energy of various clusters was calculated [23]. Linear (arching) clusters and small compact clusters were both considered, the structures of which are shown in Fig. 1.12b and c. Over the entire size range studied (up to 13 C atoms) arching clusters were found be the most stable structure. However the formation energy of the compact clusters decreases more as the size increases, so it is possible that for larger sizes these will become more stable than arching clusters for larger N.

Arching clusters have also found to be the most stable on the Ir(111) terrace for carbon clusters of up to 10 C atoms [24]. The structures for the these clusters and their formation energies are shown in Fig. 1.13b and c respectively. For Ir step edges (the (332) and (322) surfaces) the formation energies for all cluster sizes are lower, and in some cases the compact structures are more stable than the linear structures. This effect, which was also reported in [25] suggests that clusters may prefer to nucleate at step edges rather than terraces. Nevertheless the actual location of nucleation has been shown to be dependent on experimental conditions, where depositing the hydrocarbon source at low temperatures and then heating results in growth beginning on terraces, compared with depositing at high temperatures which begins on step edges [7]. This is due to the mobility of species at the different temperatures.

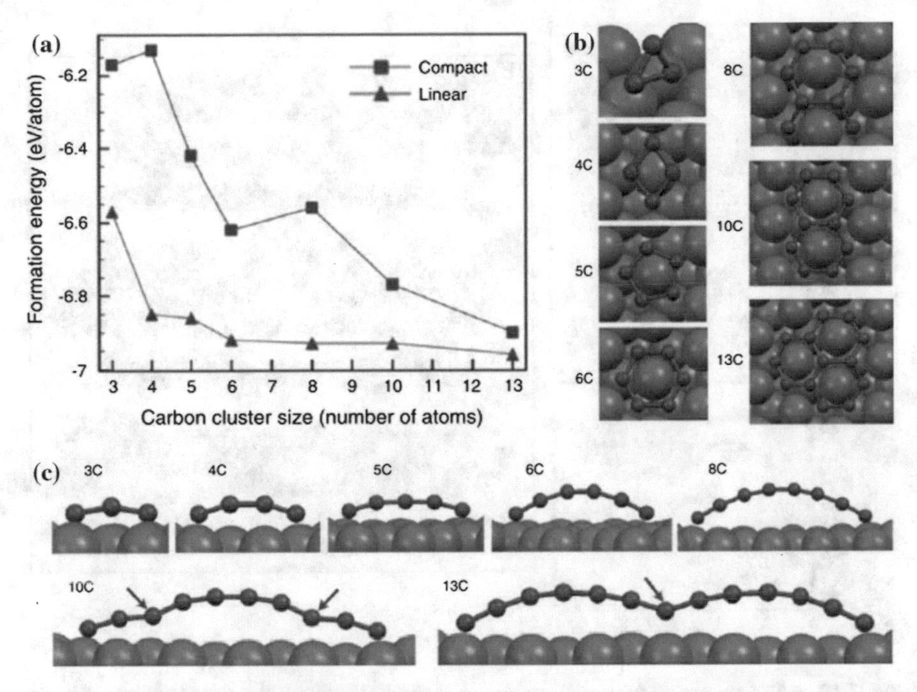

Fig. 1.12 **a** The formation energy of carbon clusters on the Cu(111) surface. Compact and linear clusters are compared, as shown in **b** and **c** respectively. Reprinted from [23], with the permission of AIP Publishing

For larger clusters ranging from 16 to 26 C atoms the formation energies of compact clusters on the Ni(111), Cu(111), Ru(0001), and Rh(111) surfaces were calculated [26]. It was found that one particular compact cluster containing 21 C atoms (shown in Fig. 1.14c) was very stable on all surfaces. These clusters were also observed experimentally on Ru(0001) and Rh(111) [19, 20].

In an extensive study of carbon clusters on nickel, the formation energy and Gibbs free energy of clusters on both the terrace Ni(111) and the step edge Ni(110) were calculated for N atoms where $N = 1 - 24$ [27]. The clusters were grouped into various types, including chain and ring-shaped structures, and sp^2 carbon networks. The formation energy for each cluster is shown in Fig. 1.15a. Overall the energy is lower for clusters at the step edges (empty shapes) than the terraces (solid shapes), and hence these clusters should be more stable. From the data the trend between cluster size and formation energy was identified for different cluster types. The formation energy of chain clusters is linearly dependent on the cluster size N, whereas for sp^2 network clusters there is a $N^{1/2}$ dependence. This means that for increasing N sp^2 networks eventually become more stable than the chain clusters. This changeover occurs for $N > 12$. By fitting curves to these formation energies (as shown in Fig. 1.15) the Gibbs free energy $G(N)$ was determined using the expression,

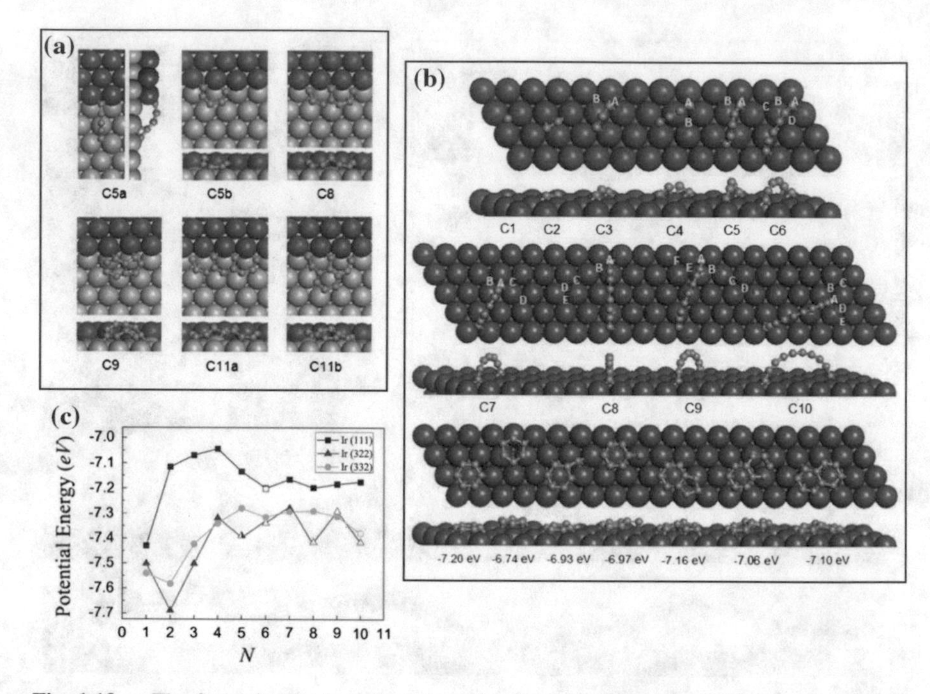

Fig. 1.13 **a** The formation (potential) energy of carbon clusters on the Ir(111) terrace, and the Ir(322) and Ir(332) step edges. **b** Linear (arching) and compact cluster structures on the terrace. For the arching clusters the most energetically favourable structures are shown. The letters indicate the positions for the end of the arch which result in other possible structures. **c** The structure of a selection of clusters at the step edge. Reprinted (adapted) with permission from [24]. Copyright (2012) American Chemical Society

$$G(N) = E(N) - \Delta\mu N. \tag{1.1}$$

where $\Delta\mu$ is the chemical potential difference between a C atom in the cluster phase and in the atomic source. This definition for the Gibbs free energy is an approximation that neglects the entropic contributions. Based on classical nucleation theory it is possible to find the nucleation barrier G^* and corresponding critical cluster size N^* from the maximum of the $G(N)$ curve. The critical cluster size N^* represents the cluster size where it is more energetically favourable for the clusters to grow in size rather than get smaller and has a nucleation barrier G^* associated with it. The dependence of G^* and N^* on the chemical potential $\Delta\mu$ is shown in Fig. 1.15b.

Both the nucleation barrier and critical cluster size are smaller for the Ni(110) surface compared to the Ni(111). Depending on the chemical potential the nucleation barrier decreases linearly from 5 to 0.2 eV between $\Delta\mu = 0.3$ and 0.8, and then is constant at 0.2 eV between 0.8 and 1. For the terrace the critical cluster size is fixed at 12 C atoms between $\Delta\mu = 0.3$ and 0.8. For these $\Delta\mu$ values carbon clusters will grow easily once the nucleation barrier is overcome at $N = 12$. For step edges the critical cluster size is reduced to $N = 10$. For both surfaces above $\Delta\mu = 0.8$

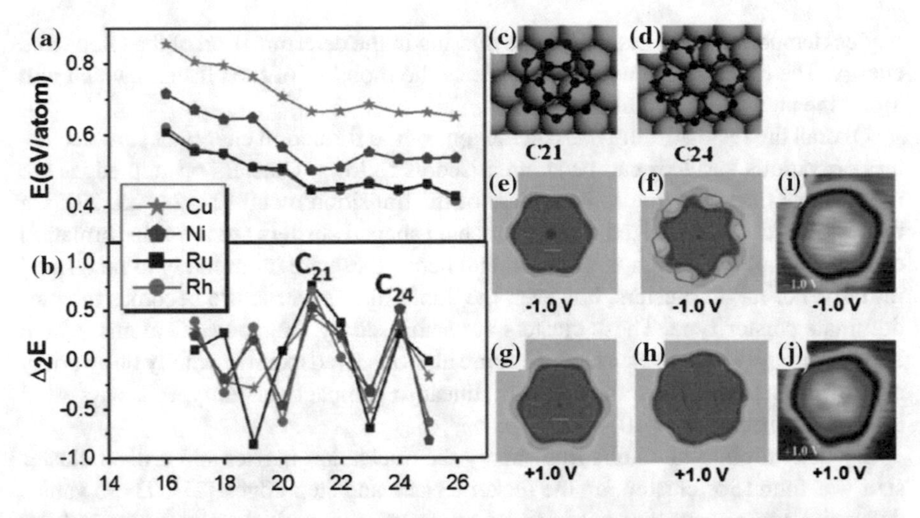

Fig. 1.14 **a** The formation energies of carbon clusters C_N on Cu(111), Ni(111), Ru(0001) and Rh(111) with N = 16, ..., 26 and **b** the second derivative of the energy with respect to N. The stable **c** C_{21} and **d** C_{24} clusters on a Rh(111) surface. Calculated STM images of the **e**, **f** C_{21} and **f**, **h** C_{24} clusters for negative and positive bias voltages, and the corresponding experimental STM images of the most abundant carbon cluster **i**, **j**. Reprinted (adapted) with permission from [26]. Copyright (2012) American Chemical Society

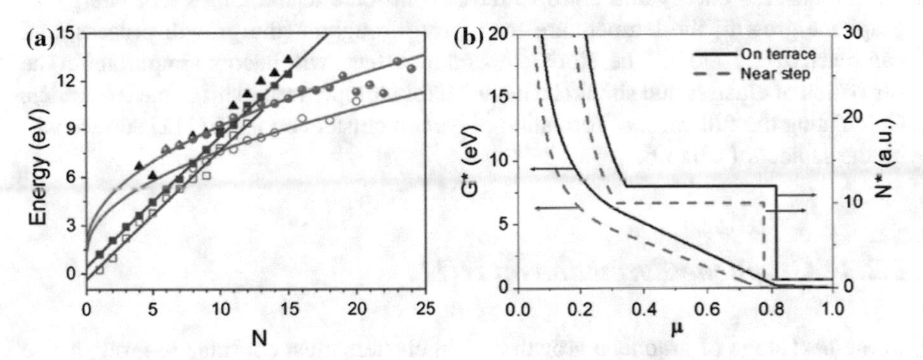

Fig. 1.15 **a** The formation energy of carbon clusters on the Ni(111) and Ni(110) surfaces, indicated by the full and empty symbols respectively. *Squares*, *triangles* and *circles* represent chains, rings and sp^2 networks respectively. **b** The nucleation barrier G* and the critical cluster size N* for clusters on the terrace and near a step edge. The curves are determined from the maximum of Eq. 1.1. Reprinted (adapted) with permission from [27]. Copyright (2012) American Chemical Society

the critical cluster size falls to 1C. This means that for a large chemical potential difference the nucleation is spontaneous. These results all suggest that nucleation will be preferable at step edges rather than terraces. However although the G* and N* are both lower at step edges than at terraces, nucleation at terraces may still be possible when the chemical potential difference is large, especially since a greater area of the nickel surface will be covered by terraces. The results of this work also neglect temperature effects, which are missing in the determination of the Gibbs free energy. The temperature will also influence the mobility of the clusters, which will affect the nucleation conditions.

Overall the results from DFT calculations of the formation energy of carbon clusters on various surfaces can be summarised as follows: Clusters on step edges are more stable than on terraces regardless of the transition metal [22, 24, 25, 27]. On the terrace, for smaller sized arching, or chain shaped clusters are lower in formation energy than all types compact cluster, and hence these are more likely to be formed initially. For larger clusters, however, the dome-like sp^2 structure becomes the predominate cluster type. These clusters are stabilised by their outer C atoms, which interact strongly with the surface. They are also observed experimentally on a variety of growth surfaces. The transition from linear to compact cluster types is suggested to occur at around $N = 10 - 12$.

By determining the Gibbs free energy the nucleation barrier and critical cluster size was found for clusters on the nickel terrace and step edges [27]. These values determine the energy that needs to be overcome for nucleation to occur and the smallest possible stable cluster size. However the method employed in this particular paper fails to include the temperature dependent effects into the Gibbs free energy, and is limited to considering a range of chemical potential differences over which G(N) is calculated. In fact for each cluster temperature dependant effects, including vibrational free energy and entropic terms contribute to the Gibbs free energy. In graphene growth, the temperature may vary throughout the growth process, and can reach over 1000 K. Therefore temperature effects will become important in the formation of clusters and should be included for a proper treatment of the energetics. Calculating the full work of formation of carbon clusters on the Ir(111) surface will be the subject of Chap. 6.

1.2.3 Graphene Formation on Ir(111)

In the next stage of graphene growth carbon clusters must continue to grow in size to form the ordered graphene structure. The manner with which the graphene forms and then grows depends on the growth conditions and the choice of substrate. This will also depend on whether there is a constant flux of growth precursors at high temperatures as in CVD, or the precursors are deposited prior to heating as in TPG. The discussion in this section is limited to the case of graphene on the Ir(111) surface which is the main interest of this Thesis.

For graphene grown on the Ir(111) surface the differences resulting from TPG and CVD growth were investigated [7]. For TPG the sample was dosed with ethylene and then heated to different temperatures for 20 seconds each. For higher heating temperatures the growth process advances further such that larger graphene islands are formed, and the density of islands is decreased. The images of the graphene produced from each heating temperature are shown in Fig. 1.16.

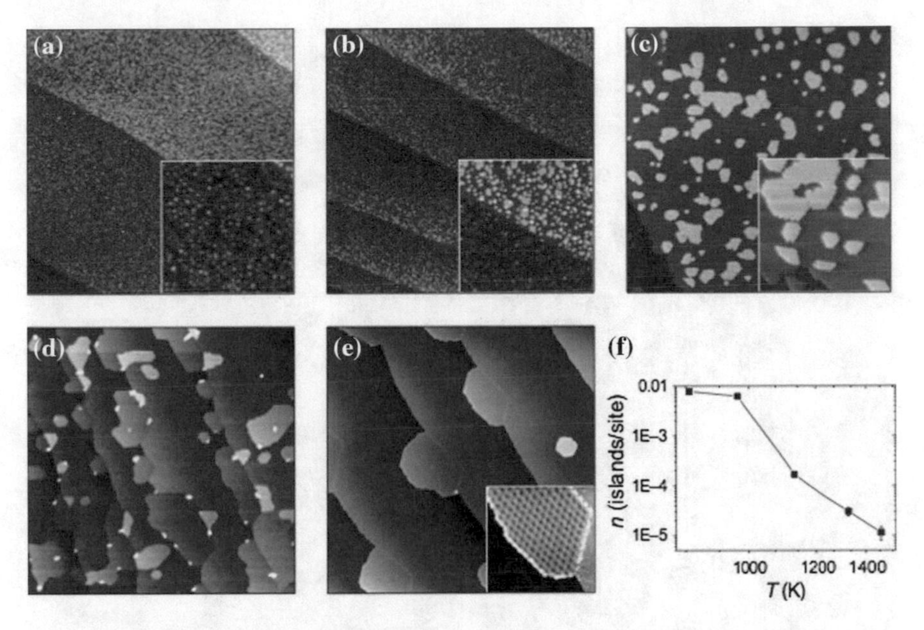

Fig. 1.16 STM images of graphene grown on the Ir(111) surface with TPG after heating to **a** 870 K **b** 970 K **c** 1120 K, **d** 1320 K, and **e** 1470 K for 20 s. **f** Graphene island density n for the different annealing temperatures T [7]

At 870 K (Fig. 1.16a) carbon species are found on the surface without the moiré pattern that is characteristic of graphene. Above 970 K graphene islands with zigzag edges are produced on both terraces and step edges. The island density decreases with temperature due to the formation of larger graphene islands as shown in Fig. 1.16f.

By imaging individual graphene islands changes in their size and shape depending on the growth temperature were also observed. Initially when graphene forms at 970 K the islands are small and compact. At 1120 K the islands become larger in size, and have a non-compact, irregular shape. At temperatures above 1320 K the islands then return to being compact, although they still continue to grow in size. These changes are thought to be the result of graphene growing via Schmoluchowski ripening. In this process islands grow by coalescing with each other, rather than by the addition of single adatoms which is a characteristic of Ostwald ripening. At 1120 K the graphene islands are mobile on the surface and can diffuse to meet and combine with each other. However there is not enough thermal energy for recently merged islands to reconstruct their shape, and hence the islands are non-compact. Above 1320 K the islands may have enough energy to return to their compact shape after coalescence.

For graphene grown with CVD the growth process is different. In this case ethylene was dosed onto the surface continuously at different temperatures. The STM images of graphene islands grown at various temperatures is shown in Fig. 1.17.

Graphene formation begins mainly at step edges. The graphene then continues to grow both up and down the step. At higher temperatures islands coalesce to form a

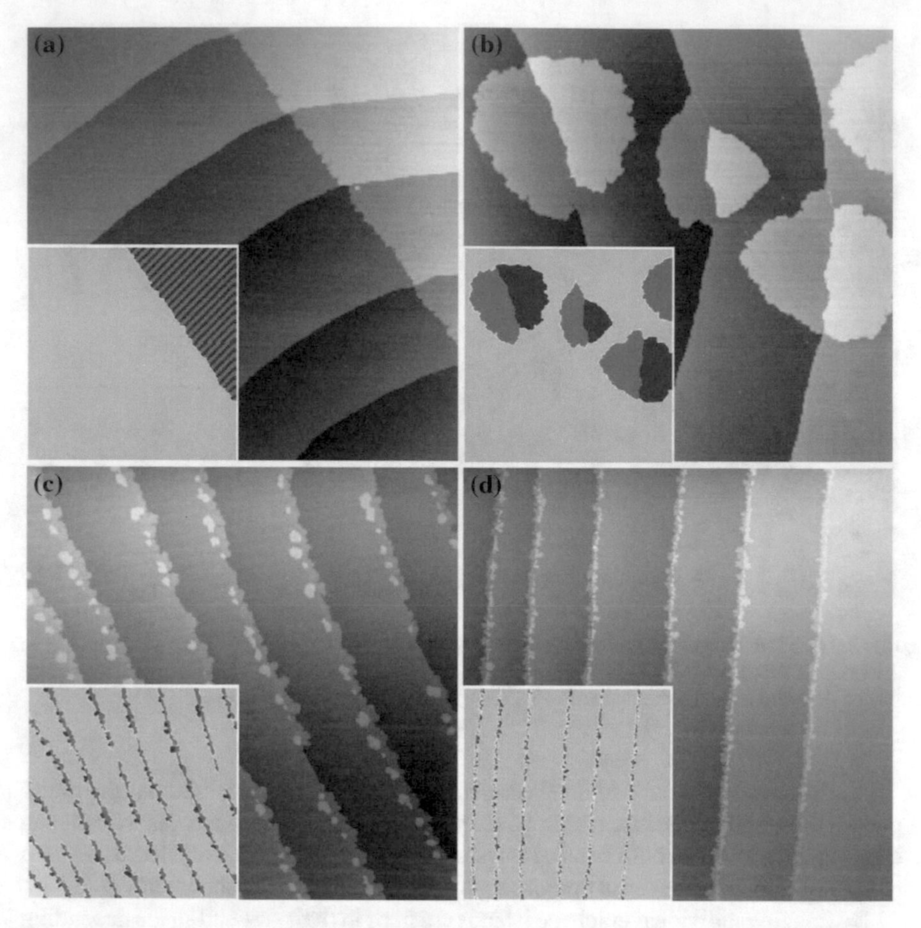

Fig. 1.17 STM images of graphene on Ir(111) during CVD growth. Images taken at 300 K after growth has been allowed to continue for: **a** 80 s and **b–d** 40 s. Surface temperatures are: **a** 1320 K, **b** 1120 K, **c** 970 K, **d** 870 K, and in all cases the pressure of ethylene during deposition was 5×10^{-10} mbar. *Insets* show graphene attached to ascending step edges (*blue*) and descending steps (*red*) [7]

continuous layer of graphene on the surface. These results suggest that nucleation occurs mainly at the step edges on the Ir(111) surface, which is also supported by the onset of growth at lower temperatures.

The differences in TPG and CVD growth are suggested to be due to the initial mobility of the carbon species [7]. In CVD the ethylene molecules arrive with some initial kinetic energy so that after their initial decomposition the resulting carbon species are very mobile, and can diffuse to the Ir step edges where they will attach with high probability before nucleation can begin. For TPG there is less thermal energy, especially at the lower growth temperatures. The mobility of species is reduced such that nucleation occurs equally on terraces and step edges. It is only at higher temperatures that the graphene islands can diffuse and become trapped at the step edges.

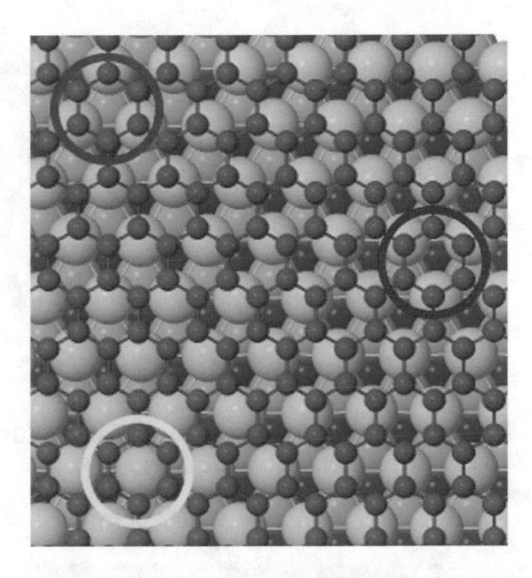

Fig. 1.18 An area of the graphene/Ir(111) moiré lattice showing the high symmetry regions. The *top* (*yellow*), hcp (*purple*) and fcc (*red*) regions are *circled* where the graphene units encircle the respective Ir site. The first three layers in the Ir(111) surface are coloured from light (*top*) to *dark blue* (third layer)

1.2.4 Graphene Substrate Interaction

Once large graphene islands have grown on a substrate they will form a moiré superstructure. This superstructure arises due to the lattice mismatch between the graphene and the substrate. The superstructure can develop periodic characteristics depending on the strength of the interaction of the graphene with the substrate.

The size of the moiré lattice is defined depending on the number of graphene and metal atoms that build up the repeating unit cell. The structure produced has a periodicity derived from the superposition of graphene hexagon units ($m_1 \times m_2$) on surface cells ($n_1 \times n_2$). Depending on substrate, and its lattice constant, many graphene units may be required to build up the unit cell. Across the moiré superstructure different regions formed depending on how the carbon atoms in the graphene are arranged with respect to the metal atoms. For example for the Ir(111) surface hcp, fcc and top regions exist where the graphene hexagon units line up above these sites on the surface. This is shown in Fig. 1.18.

The position of the individual carbon atoms with respect to the metal atoms will also affect the interaction between the graphene and the surface, where C atoms positioned directly on top of the metal atoms being more strongly bound. The distribution of regions where the graphene is bound more strongly (where C atoms closer to the surface) and less strongly (where the C atoms are further away) results in periodic corrugations in the graphene structure. For metals with a strong interaction with graphene large corrugations are formed. This is the case for Ru(0001) where the height of the corrugations were measured with low energy electron diffraction (LEED) as being 1.5 Å [28]. An illustration of these corrugations is shown in Fig. 1.19. For graphene on Ir(111) (in the R0 phase) the reduced interaction gives smaller corrugations with a height of 0.3 Å measured in [29, 30].

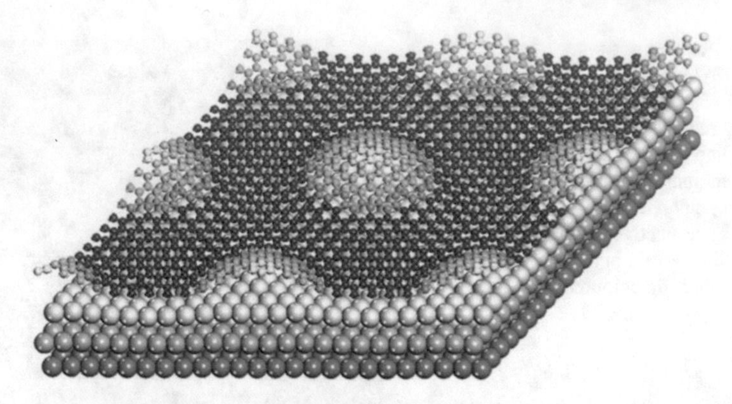

Fig. 1.19 A 3D model of graphene on the Ru(0001) surface showing the periodic corrugations (determined from LEED results). Reprinted figure with permission from [[28] as follows: W. Moritz, B. Wang, M.-L. Bocquet, T. Brugger, T. Greber, J. Wintterlin, and S. Günther, Physical Review Letters, 104, 146102, 2010] Copyright (2010) by the American Physical Society

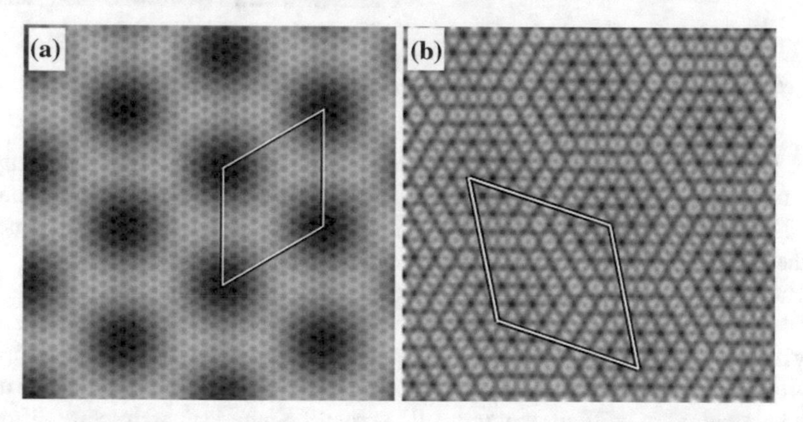

Fig. 1.20 STM images of **a** the R0 and **b** the R30 domains of graphene on Ir(111). Reprinted figure with permission from [30] as follows: E. Loginova, S. Nie, K. Thürmer, N.C. Bartelt, and K.F. McCarty, Phys. Rev. B, 80, 085430, 2009. Copyright (2009) by the American Physical Society

Different orientations of the graphene with respect to the substrate are also possible. For metals which strongly interact with graphene such as Ru(0001) and Rh(111) only the R0 phase is found [28, 31]. In this case the graphene lattice is aligned with respect to the substrate lattice. Other metals only weakly interact with graphene via Van der Waals forces. In this case multiple rotational domains are possible. For the Ir(111) surface the R30 domain is observed [30] in addition to the R0 domain as shown in Fig. 1.20, and many others are also predicted with DFT [32]. Each domain has its own moiré structure and therefore has different corrugation. Experimental and theoretical results for the properties of the moiré structure of graphene on the Ru(0001), Rh(111), Ir(111) and Pt(111) surfaces and listed in Table 1.2. The experimental data is usually derived from STM, LEED and atomic force microscopy (AFM) experiments.

Table 1.2 The details of the structure of graphene on various surfaces. The surfaces are listed in the order from the strongest graphene-surface interaction (Ru(0001)) to the weakest (Pt(111)) [40]. The corrugation is described in terms of the peak-to peak height with the data in brackets being from DFT calculations

Surface	Lattice constant Å	Angle of graphene to surface	Graphene moiré superstructure	Corrugation Å
Ru(0001)	2.71	0°	(25 × 25) [28, 33]	0.82 [33], 1.53 [28], 0.15 [35] (1.5 [36])
			(13 × 13) [28]	(1.59 [28])
			(12 × 12) [34]	(1.5 [34])
Rh(111)	2.69	0°	(12 × 12) [31]	0.5–1.5 [31]
Ir(111)	2.72	0° [30, 32]	incommensurate [37], (10 × 10)[32]	~ 0.3 [29, 30] (0.423 [32])
		~14 ° [30, 32]	(4 × 4) [32]	(0.101 [32])
		19° [32]	(3 × 3) [32]	(0.051 [32])
		23° [32]	$(\sqrt{19} \times \sqrt{19})$ [32]	(0.022 [32])
		26° [32]	$(\sqrt{37} \times \sqrt{37})$ [32]	(0.015 [32])
		30° [30, 32]	(2 × 2) [32]	0.04 [30] (0.014 [32])
Pt(111)	2.77	30°	(2 × 2)	
		19°	(3 × 3)	<0.3
		14°	(4 × 4)	
		[38]	[38, 39]	[38]
		6°	$(\sqrt{37} \times \sqrt{37})$	
		3°	$(\sqrt{61} \times \sqrt{61})$	0.5–0.8
		2°	$(\sqrt{67} \times \sqrt{67})$	

Overall the results in Table 1.2 show the importance of the graphene/substrate interaction. The corrugation varies greatly depending on the surface, and/or orientation of the graphene to the substrate. Such effects will influence graphene growth, and the electronic structure once graphene has formed.

1.2.5 Removing Defects

After the epitaxial growth process the graphene produced is unlikely to have a perfect structure. As well as the graphene/substrate interaction that warps the graphene sheet, the graphene produced will likely contain many defects in its structure. These are formed during the growth process. These defects include grain boundaries, and point defects such as vacancy, interstitial and substitutional defects. Grain boundaries are formed where two growth fronts meet to form a discontinuity in the growth. Vacancy defects form when there are missing atoms in the structure, whereas interstitial defects involve extra atoms. Substitutional defects are produced when a carbon atom is replaced with another type, usually the same as the substrate.

Overall defects negatively affect the properties of graphene, reducing its carrier mobility that is important in many of its potential applications. By carefully controlling the growth domains it is possible to reduce the grain boundaries. However it has been shown that these alone cannot account for the total loss in carrier mobility. Therefore point defects must contribute to the reduced quality of graphene [41, 42]. It is therefore important to understand the nature of these defects, and whether it is possible to remove them from the graphene structure once they have formed.

It was suggested in [8] that point defects in graphene can be healed once formed during epitaxial growth. Large strain fields were found in graphene grown with CVD on the Ir(111) surface. The presence of strain in the system is demonstrated through changes in the graphene lattice parameter, where large strain fields correspond to a larger average lattice parameter. These effects were both attributed to the presence of point defects in the structure. By dosing the system with ethylene at high temperatures it was found that the graphene lattice parameter decreases as the overall graphene coverage increases. The reflection high-energy electron diffraction (RHEED) technique was used to determine the diffraction pattern for the change in the graphene lattice parameter (a_c), and the change in a_c with respect to the coverage. These are shown in Fig. 1.21a and b respectively. The coverage increases with the ethylene dosage.

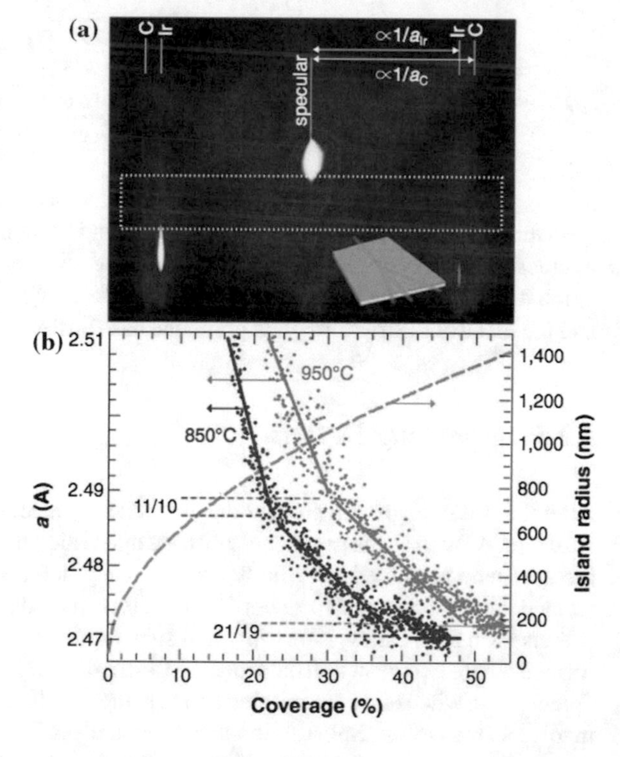

Fig. 1.21 **a** The RHEED pattern for graphene on the Ir(111) surface, where the distance between the central streak and the outer streaks is inversely proportional to the graphene lattice constant and the Ir lattice constant. **b** The change in the graphene lattice parameter and the island radius as a function of coverage at 850 and 950° C. Reprinted figure with permission from as follows: [[8] N. Blanc, F. Jean, A.V. Krasheninnikov, G. Renaud, and J. Coraux, Phys. Rev. Lett., 111, 085501, 2013.] Copyright (2013) by the American Physical Society

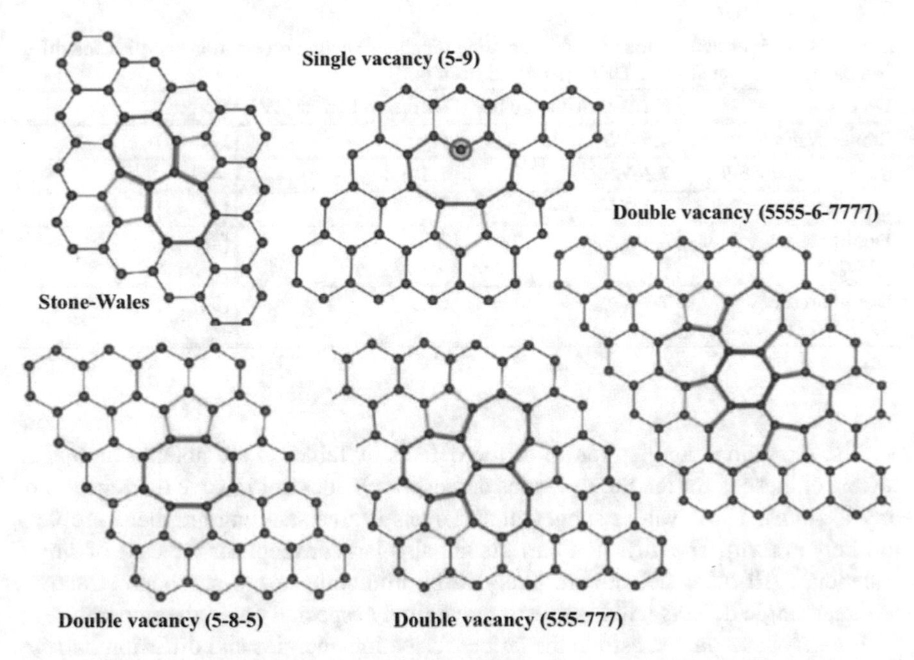

Fig. 1.22 The structure of some of the point defects found in free-standing graphene. Reprinted (adapted) with permission from [43]. Copyright (2011) American Chemical Society

Figure 1.21b also shows a series of phase changes in the graphene lattice parameter. These are suggested to be due to the formation of different moiré superstructures as the tensile strain decreases. As vacancies are healed the commensurate phase changes from (11 × 11)/(10 × 10) to (21 × 21)/(19 × 19). The 11/10 phase is thought to maximise the graphene/iridium interaction, which helps to stabilise the strained system. Overall the changes in the strain and a_c are suggested to be due to the healing of the vacancy defects. The proposed mechanism for this is that defects diffuse to the edge of the graphene where they can be removed from the structure. However this mechanism was not actually verified theoretically.

In order to understand how vacancy defect healing might be possible it is important to understand the structure and energetics of the different types of defect. This can be achieved using DFT calculations. For defects in graphene in the gas phase the situation is much simpler than if the graphene is on top of a substrate. All interactions with the substrate, and effects due to the moiré superstructure can be ignored.

A detailed review of the point defects in free-standing graphene was reported in [43]. Images of some of the different types of defect are shown in Fig. 1.22.

Stone Wales defects are formed when a C-C group of the graphene is rotated by 90° in the structure. This results in four graphene hexagon units being replaced by two pentagons and two heptagons. For a single vacancy a 5–9 structure is formed, whereas for a double vacancy a variety of shapes are possible. A 5-8-5, 555–777 or a 5555-6-7777 reconstruction may be formed as shown in the figure. The calculated formation energies and diffusion barriers for these defects are shown in Table 1.3.

Table 1.3 Various defects found in free standing graphene and their formation energies and diffusion barriers calculated using DFT. [Adapted from [43]]

Defect	Formation energy [eV]	Diffusion barrier [eV]	Reference
Stone–Wales **55-77**	4.5–5.3	10	[44, 45]
Single vacancy **5-9**	7.3–7.5	1.2–1.4	[46]
Double vacancy **5-8-5**	7.2–7.9	7	[46, 47]
Double vacancy **555-777**	6.4–7.5	6	[48, 49]
Double vacancy **5555-6-7777**	7	6	[50, 51]

The formation energies for all of the defects in Table 1.3 are notably high. The lowest of these is for the Stone–Wales defect which does not involve the removal of any C atoms. These values suggest that defects in free standing graphene are very unlikely to form. The diffusion barriers are also large, except for the case of single vacancies. All other defects are likely to be immobile in the graphene structure, whereas single defects will be able to move freely especially at high temperatures.

For graphene on a substrate the defect formation energies and diffusion barriers will naturally be affected by the interaction with the substrate. The presence of the underlying substrate also means that more defect structures are possible since the defective graphene can bond with the substrate, and incorporate atoms from the substrate into its structure.

In a theoretical paper the structure and energetics of single and double vacancy defects on the Cu(111), Co(0001) and Ni(111) surfaces were studied [52]. For these defects the possible structures also include the defect site being filled by a metal atom from the substrate (3DBs and 4DBs) and an additional metal atom filling the defect site (M@3DBs and M@4DBs), where the number preceding the DB indicates the number of dangling bonds. The structures of all these defects are shown in Fig. 1.23 and their formation energies are shown in Fig. 1.24.

Compared to free-standing graphene it was found that defects on metal surfaces have a lower formation energy, due to the interaction with the metal surface. However the formation energies are still fairly large at 2.5–7 eV. For 3DBs single vacancies the formation energy is the lowest at under 3 eV on the Ni(111) and Co(0001) surfaces. For vacancies on the Cu(111) surface the formation energy is higher than on the other surfaces for all defect structures. These values suggest that graphene grown on the Cu(111) surface may contain less defects than on the Ni(111) or the Co(0001) surfaces. This is probably due to the weaker interaction between copper and graphene.

The formation energy of single vacancy defects in graphene on the Ir(111) surface were also calculated in [53]. Here the whole moiré structure of graphene on Ir(111) was taken into account (see Sect. 1.2.4). Across the moiré lattice defects can be formed in the high symmetry regions, such as the top, fcc and hcp regions. These are shown in Fig. 1.25. The position of the defect site with respect to the iridium

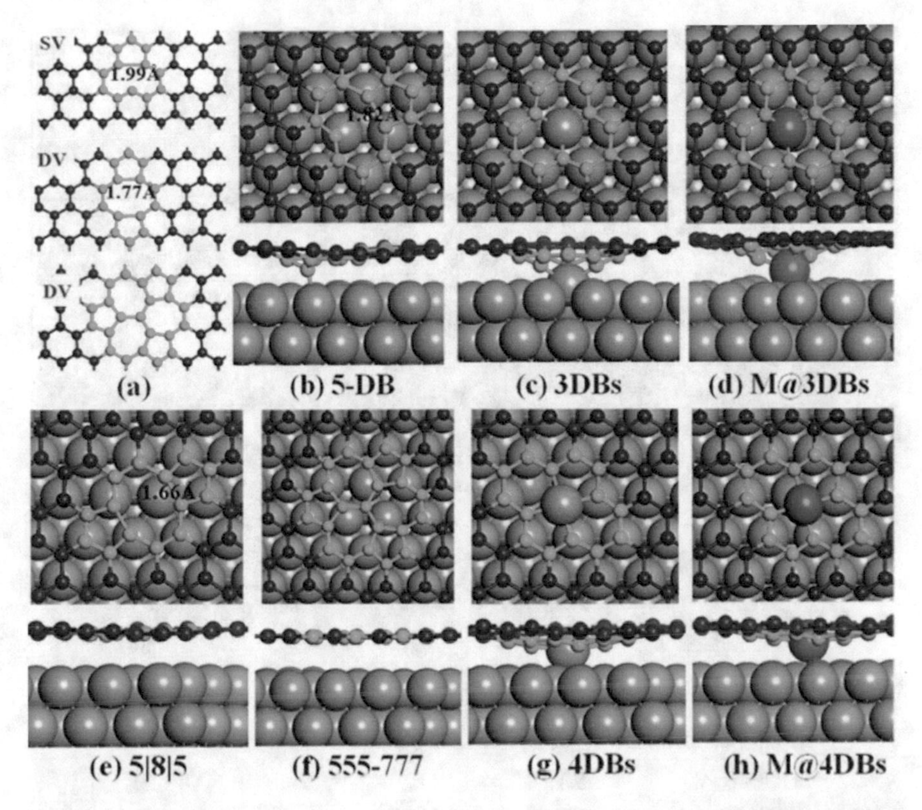

Fig. 1.23 **a** Optimised structures of a single vacancy and the 5-8-5 and 555-777 double vacancies in free-standing graphene. The single vacancy structures on the Cu(111) surface: **b** 5-DBs as in free standing graphene, **c** 3DBs and **d** M@3DBs, as well as the double vacancy structures **e** 5-8-5, **f** 555-777, **g** 4DBs and **h** M@4DBs. The carbon atoms around the defect and the extra metal adatom in the M@3DBs and M@4DBs structures are *coloured green* and *blue*, respectively. Reprinted (adapted) with permission from [52]. Copyright (2013) American Chemical Society

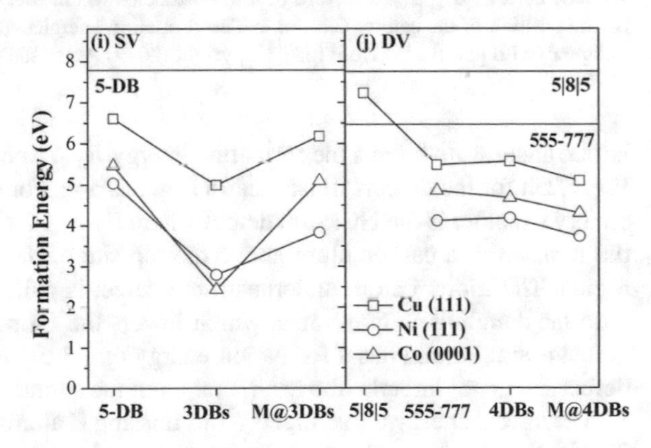

Fig. 1.24 The formation energies of the various single and double vacancy defects as shown in Fig. 1.23. These value were calculated using the C chemical potential from free- standing graphene, and the chemical potential of the metal atom from the bulk. Reprinted (adapted) with permission from [52]. Copyright (2013) American Chemical Society

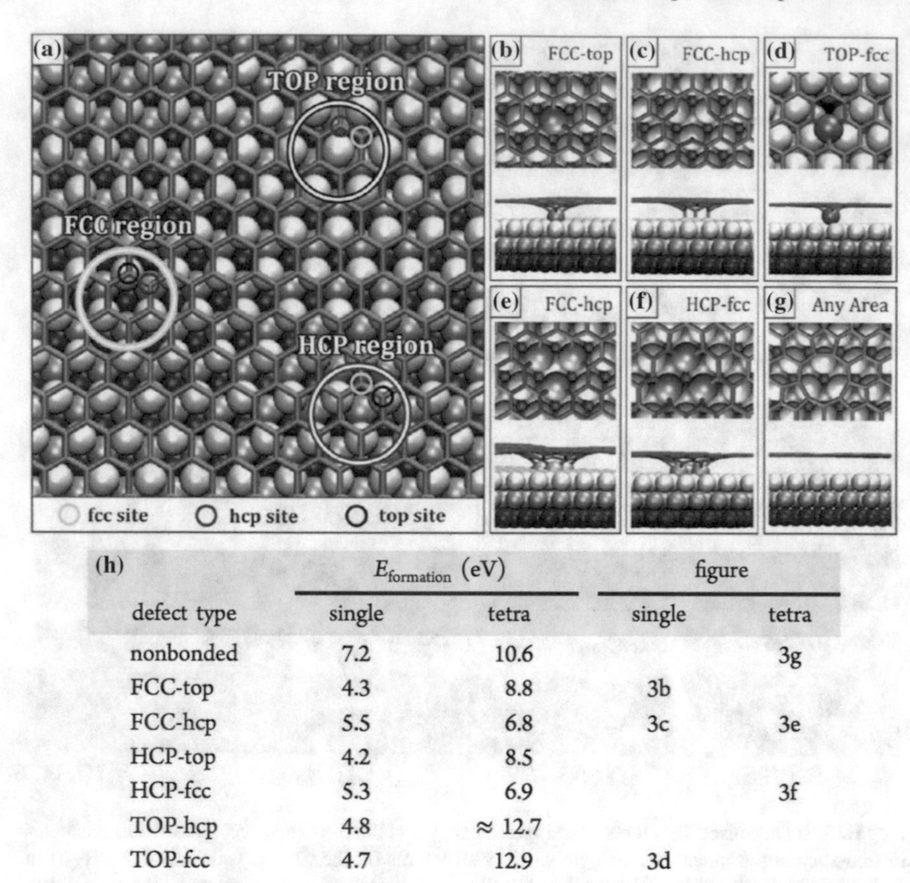

	$E_{formation}$ (eV)		figure	
defect type	single	tetra	single	tetra
nonbonded	7.2	10.6		3g
FCC-top	4.3	8.8	3b	
FCC-hcp	5.5	6.8	3c	3e
HCP-top	4.2	8.5		
HCP-fcc	5.3	6.9		3f
TOP-hcp	4.8	≈ 12.7		
TOP-fcc	4.7	12.9	3d	

Fig. 1.25 **a** Different regions graphene on the Ir(111) surface as indicated by where the graphene units enclose the hcp, fcc or top sites of the surface (upper case). The defect positions are related to which site the removed C atom is above (lower case). **b–d** The structure of various single vacancy defects. **e–g** The structure of tetra-vacancies where four C atoms are removed, as given by the position of the central C atom. **h** The formation energies of the various defects. Reprinted (adapted) with permission from [53]. Copyright (2013) American Chemical Society

lattice underneath affects the formation energy E_f of the defect. These are shown in Fig. 1.25h for the defects illustrated in Fig. 1.25a–g. For single vacancies the lowest energy structure is the HCP-top defect where $E_f = 4.2$ eV. This defect is formed by the removal of a carbon atom above the top site of the Ir(111) surface, in the hcp region. The iridium atom underneath the defect is pulled up and interacts strongly with the dangling carbon bonds which lowers the formation energy. The FCC-top defect is similar and has a formation energy of 4.3 eV. For other defects where the defect site is not directly above a metal atom the formation energy is higher.

The lowest energy tetra defects (four missing C atoms) are similar in structure to the single vacancies, where strong interactions between the dangling carbon bonds and the iridium surface atoms help to stabilise the structure.

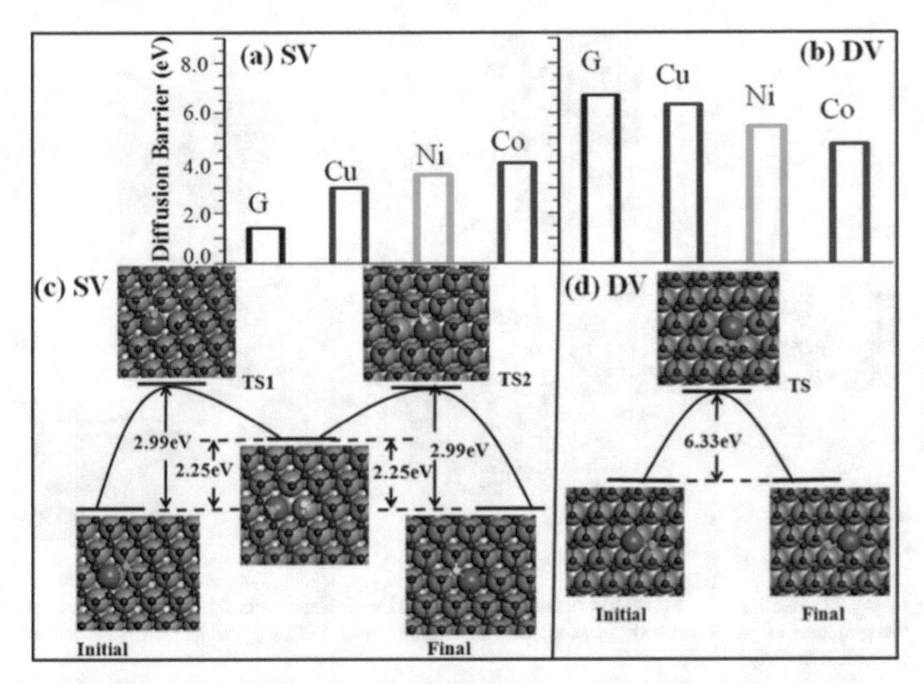

Fig. 1.26 The diffusion barriers of **a** the M@3DBs single vacancy and **b** the M@4DBs double vacancy on the Cu(111), Co(0001) and Ni(111) surfaces and in free standing graphene (G). **c** and **d** show the diffusion pathways for the single and double vacancies on Cu(111) respectively. Reprinted (adapted) with permission from [52]. Copyright (2013) American Chemical Society

Defect diffusion was also considered [52]. This is important because defects that are mobile in the graphene structure may exist on shorter timescales since they are able to diffuse to the edge of the graphene, or meet and coalesce with other defects. For the most stable single vacancy and double vacancy structures the diffusion barriers were calculated [52]. The values, along with the diffusion mechanisms are shown in Fig. 1.26. For the single vacancy (3DBs) and the double vacancy (M@4DBs) on the Cu(111) surface the barriers were calculated as 2.99 and 6.33 eV respectively. Overall the defects are concluded to mostly be immobile, due to their high diffusion barriers. For single vacancies on Cu(111) however diffusion may take place, albeit slowly (about 10 nm per hour). However as a result of this double vacancies may be formed more easily if two single vacancies are able to diffuse to meet each other.

The removal of metal atoms at the edge of a graphene front was also investigated. To remove the metal atoms two carbon atoms are required to diffuse to the edge and then the suggested process occurs in two steps. First one carbon atom must attach onto the metal atom to form a hexagon, and then the next carbon atom replaces the metal to complete the graphene edge. The whole process is depicted in Fig. 1.27 for graphene on the Cu(111) surface. The barriers in this case are 1.86 eV for the first transition state and 0.33 eV for the second transition state.

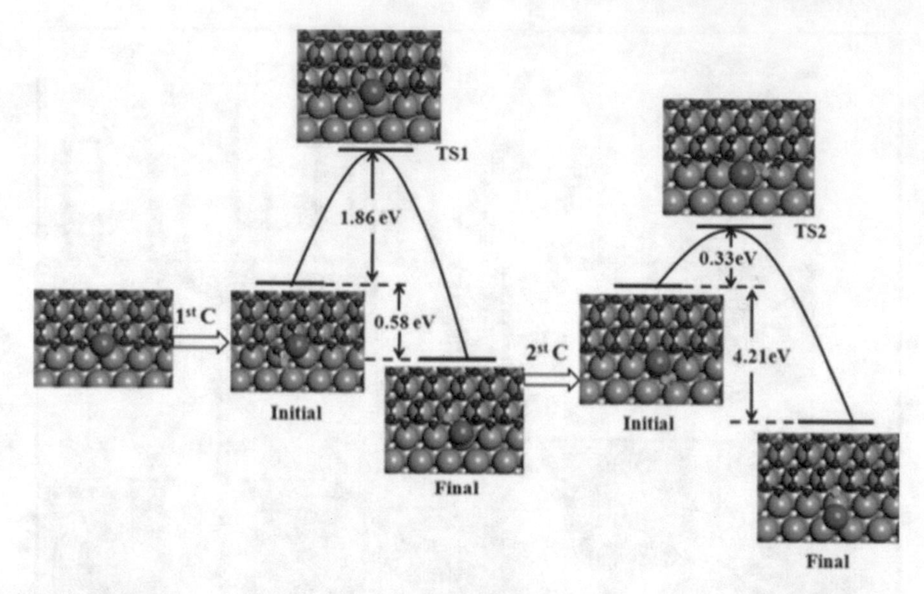

Fig. 1.27 Healing of a M@4DBs defect on the Cu(111) surface via its replacement with C at the graphene edge. Reprinted (adapted) with permission from [52]. Copyright (2013) American Chemical Society

The barriers for healing with this mechanism are lower than the barrier for defect diffusion and therefore this process will be limited by the amount of defects reaching the graphene edge. For single vacancies diffusion to the edge may be possible on some surfaces, however double vacancies have so far been shown to be immobile. Therefore once double vacancies are trapped in the graphene structure they will unable to heal using this mechanism.

So far there has been very limited work on the healing of graphene defects, where graphene is on a substrate, as in epitaxial growth. Although the structure and formation energies of various vacancy defects on different transition metal surfaces have been calculated there has been very little work considering the kinetics of defects. This may be in part due to the complicated structure of defects in graphene/substrate systems, which requires a large unit cell in order to capture effects from the moiré superstructure.

If healing of vacancy defects is possible as reported in [8] it must be feasible for them to heal without diffusing to the graphene edge front, since only single vacancies may be mobile. Therefore the direct healing of defects within the graphene sheet will be investigated by depositing ethylene molecules directly towards the defects, under high temperatures. Molecular dynamics simulations and NEB calculations of this healing process will be the subject of Chap. 7.

References

1. A. Hudson, *Is Graphene a Miracle Material?* http://news.bbc.co.uk/1/hi/programmes/click_online/9491789.stm (May 21, 2011)
2. S.V. Morozov, D. Jiang, M.I. Katsnelson, I.V. Grigorieva, S.V. Dubonos, K.S. Novoselov, A.K. Geim, A.A. Firsov, Two-dimensional gas of massless dirac fermions in graphene. Nature **438**, 197–200 (2005)
3. F. Schedin, T.J. Booth, V.V. Khotkevich, S.V. Morozov, K.S. Novoselov, D. Jiang, A.K. Geim, Two-dimensional atomic crystals. Proc. Natl. Acad. Sci. U.S.A. **102**(30), 10451–10453 (2005)
4. A.A. Balandin, S. Ghosh, W. Bao, I. Calizo, D. Teweldebrhan, F. Miao, C.N. Lau, Superior thermal conductivity of single-layer graphene. Nano Lett. **8**(3), 902–907 (2008)
5. C. Lee, X. Wei, J.W. Kysar, J. Hone, Measurement of the elastic properties and intrinsic strength of monolayer graphene. Science **321**(5887), 385–388 (2008)
6. A.K. Geim, K.S. Novoselov, The rise of graphene. Nature Mater. **6**, 183–191 (2007)
7. J. Coraux, A.T. N'Diaye, M. Engler, C. Busse, D. Wall, N. Buckanie, F.-J. M. zu. Heringdorf, R. van Gastel, B. Poelsema, T. Michely, Growth of graphene on Ir(111). New J. Phys. **11**(023006) (2009)
8. N. Blanc, F. Jean, A.V. Krasheninnikov, G. Renaud, J. Coraux, Strains induced by point defects in graphene on a metal. Phys. Rev.Lett. **111**, 085501 (2013)
9. H. Tetlow, J. Posthuma de Boer, I.J. Ford, D.D. Vvedensky, J. Coraux, L. Kantorovich, Growth of epitaxial graphene: theory and experiment. Phys. Rep. **542**(3), 195–295 (2014)
10. Z. Li, P. Wu, C. Wang, X. Fan, W. Zhang, X. Zhai, C. Zeng, Z. Li, J. Yang, J. Hou, Low-temperature growth of graphene by chemical vapor deposition using solid and liquid carbon sources. ACS Nano **5**(4), 3385–3390 (2011)
11. T. Fuhrmann, M. Kinne, B. Tränkenschuh, C. Papp, J.F. Zhu, R. Denecke, H.-P. Steinrück, Activated adsorption of methane on Pt(111) - an in situ XPS study. New J. Phys. **7**, 107–107 (2005)
12. S. Lizzit, A. Baraldi, High-resolution fast X-ray photoelectron spectroscopy study of ethylene interaction with Ir(111): From chemisorption to dissociation and graphene formation. Catal. Today **154**, 68–74 (2010)
13. W. Zhang, P. Wu, Z. Li, J. Yang, First-principles thermodynamics of graphene growth on Cu surfaces. J. Phys. Chem. C **115**(36), 17782–17787 (2011)
14. G. Gajewski, C.W. Pao, Ab initio calculations of the reaction pathways for methane decomposition over the Cu (111) surface. J. Phys. Chem. C **135**(6), 064707 (2011)
15. M. Li, W. Guo, R. Jiang, L. Zhao, X. Lu, H. Zhu, D. Fu, H. Shan, Mechanism of the ethylene conversion to ethylidyne on Rh(111): a density functional investigation. J. Phys. Chem. C **114**(18), 8440–8448 (2010)
16. Z.-J. Zhao, L.V. Moskaleva, H.A. Aleksandrov, D. Basaran, N. Rösch, Ethylidyne formation from ethylene over Pt(111): a mechanistic study from first-principle calculations. J. Phys. Chem. C **114**(28), 12190–12201 (2010)
17. H.A. Aleksandrov, L.V. Moskaleva, Z.-J. Zhao, D. Basaran, Z.-X. Chen, D. Mei, N. Rösch, Ethylene conversion to ethylidyne on Pd(111) and Pt(111): A first-principles-based kinetic Monte Carlo study. J. Catal. **285**(1), 187–195 (2012)
18. Y. Chen, D.G. Vlachos, Hydrogenation of ethylene and dehydrogenation and hydrogenolysis of ethane on Pt(111) and Pt(211): a density functional theory study. J. Phys. Chem. C **114**(11), 4973–4982 (2010)
19. B. Wang, X. Ma, M. Caffio, R. Schaub, W.-X. Li, Size-selective carbon nanoclusters as precursors to the growth of epitaxial graphene. Nano Lett. **11**, 424–430 (2011)
20. Y. Cui, Q. Fu, H. Zhang, X. Bao, Formation of identical-size graphene nanoclusters on Ru(0001). Chem. Commun. **47**, 1470–1472 (2011)
21. P. Lacovig, M. Pozzo, D. Alfè, P. Vilmercati, A. Baraldi, S. Lizzit, Growth of dome-shaped carbon nanoislands on Ir(111): The intermediate between carbidic clusters and quasi-free-standing graphene. Phys. Rev. Lett. **103**(166101) (2009)

22. H. Chen, W. Zhu, Z. Zhang, Contrasting behaviour of carbon nucleation in the initial stages of graphene epitaxial growth on stepped metal surfaces. Phys. Rev. Lett. **104**(186101) (2010)
23. R.G. Van Wesep, H. Chen, W. Zhu, Z. Zhang, Communication: Stable carbon nanoarches in the initial stages of epitaxial growth of graphene on Cu(111). J. Chem. Phys. **134**(17), 171105 (2011)
24. P. Wu, H. Jiang, W. Zhang, Z. Li, Z. Hou, J. Yang, Lattice mismatch induced nonlinear growth of graphene. J. Am. Chem. Soc. **134**, 6045–6051 (2012)
25. C. Herbig, E.H. Åhlgren, W. Jolie, C. Busse, J. Kotakoski, A.V. Krasheninnikov, T. Michely, Interfacial carbon nanoplatelet formation by ion irradiation of graphene on iridium(111). ACS Nano **8**(12), 12208–12218 (2014)
26. Q. Yuan, J. Gao, H. Shu, J. Zhao, X. Chen, F. Ding, Magic carbon clusters in the chemical vapor deposition growth of graphene. J. Am. Chem. Soc. **134**, 2970 (2012)
27. J. Gao, J. Yip, J. Zhao, B.I. Yakobson, F. Ding, Graphene nucleation on transition metal surface: structure transformation and role of the metal step edge. J. Am. Chem. Soc. **133**(13), 5009–5015 (2011)
28. W. Moritz, B. Wang, M.-L. Bocquet, T. Brugger, T. Greber, J. Wintterlin, S Günther, Structure determination of the coincidence phase of graphene on Ru(0001). Phys. Rev. Lett. **104**(136102) (2010)
29. A.T. N'Diaye, S. Bleikamp, P.J. Feibelman, T. Michely, Two-dimensional Ir cluster lattice on a graphene Moiré on Ir(111). Phys. Rev. Lett. **97**, 215501 (2006)
30. E. Loginova, S. Nie, K. Thürmer, N.C. Bartelt, K.F. McCarty, Defects of graphene on Ir(111): rotational domains and ridges. Phys. Rev. B **80**, 085430 (2009)
31. E.N. Voloshina, Yu.S. Dedkov, S. Torbrogge, A. Thissen, M. Fonin, Graphene on Rh(111): scanning tunneling and atomic force microscopies studies. Appl. Phys. Lett. **100**(241606) (2012)
32. L. Meng, R. Wu, L. Zhang, L. Li, S. Du, Y. Wang, H.-J. Gao, Multi-oriented moire superstructures of graphene on Ir(111): experimental observations and theoretical models. J. Phys. Conden. Matter **24**, 314214 (2012)
33. D. Martoccia, M. Björck, C.M. Schleptz, T. Brugger, S.A. Pauli, B.D. Patterson, T. Greber, P.R. Willmott, Graphene on Ru(0001): a corrugated and chiral structure. New J. Phys. **12**(043028) (2010)
34. B. Wang, M.-L. Bocquet, S. Marchini, S. Gunther, J. Wintterlin, Chemical origin of a graphene moire overlayer on Ru(0001). J. Phys. Chem. Chem. **10**, 3530–3534 (2008)
35. B. Borca, S. Barja, M. Garnica, M. Minniti, A. Politano, J.M. Rodriguez-Garcia, J.J. Hinarejos, D. Farias, A.L. Vazquez de Parga, R. Miranda, Electronic and geometric corrugation of periodically rippled, Self-nanostructured graphene epitaxially grown on Ru(0001). New J. Phys. **12**, 093018 (2010)
36. D. Martoccia, P.-R. Willmott, T. Brugger, M. Björck, S. Günther, C.-M. Schlepütz, A. Cervellino, S.-A. Pauli, B.-D. Patterson, S. Marchini, J. Wintterlin, W. Moritz, T. Greber, Graphene on Ru(0001): A 25 x 25 supercell. Phys. Rev. Lett. **101**(126102) (2008)
37. N. Blanc, J. Coraux, C. Vo-Van, A.T. N'Diaye, O. Geaymond, G. Renaud, Local deformations and incommensurability of high-quality epitaxial graphene on a weakly interacting transition metal. Phys. Rev. B **86**, 235439 (2012)
38. M. Gao, Y. Pan, L. Huang, H. Hu, L.Z. Zhang, H.M. Guo, S.X. Du, H.-J. Gao, Epitaxial growth and structural property of graphene on Pt(111). Appl. Phys. Lett. **98**, 033101 (2011)
39. P. Merino, M. Svec, A.L. Pinardi, G. Otero, J.A. Martin-Gago, Strain-driven moire superstructures of epitaxial graphene on transition metal surfaces. ACS Nano **5**(7), 5627–5634 (2011)
40. A.B. Preobrajenski, M.L. Ng, A.S. Vinogradov, N. Mårtensson, Controlling graphene corrugation on lattice-mismatched substrates. Phys. Rev. B **78**, 073401 (2008)
41. X. Li, C.W. Magnuson, A. Venugopal, J. An, J.W. Suk, B. Han, M. Borysiak, W. Cai, A. Velamakanni, Y. Zhu, L. Fu, E.M. Vogel, E. Voelkl, L. Colombo, R.S. Ruoff, Graphene films with large domain size by a two-step chemical vapor deposition process. Nano Lett. **10**(11), 4328–4334 (2010)

42. Y. Hernandez, V. Nicolosi, M. Lotya, F.M. Blighe, Z. Sun, S. de, I.T. McGovern, B. Holland, M. Byrne, Y.K. Gun'ko, J.J. Boland, P. Niraj, G. Duesberg, S. Krishnamurthy, R. Goodhue, J. Hutchison, V. Scardaci, A.C. Ferrari, J. N. Coleman, High-yield production of graphene by liquid-phase exfoliation of graphite. Nat. Nanotechnol. **3**, 563–568 (2008)
43. F. Banhart, J. Kotakoski, A.V. Krasheninnikov, Structural defects in graphene. ACS Nano **5**(1), 26–41 (2011)
44. L. Li, S. Reich, J. Robertson, Defect energies of graphite: density-functional calculations. Phys. Rev. B **72**, 184109 (2005)
45. J. Ma, D. Alfè, A. Michaelides, E. Wang, Stone-wales defects in graphene and other planar sp^2-bonded materials. Phys. Rev. B **80**, 033407 (2009)
46. A.A. El-Barbary, R.H. Telling, C.P. Ewels, M.I. Heggie, P.R. Briddon, Structure and energetics of the vacancy in graphite. Phys. Rev. B **68**, 144107 (2003)
47. A.V. Krasheninnikov, P.O. Lehtinen, A.S. Foster, R.M. Nieminen, Bending the rules: contrasting vacancy energetics and migration in graphite and carbon nanotubes. Chem. Phys. Lett. **418**, 132–136 (2006)
48. G.-D. Lee, C.Z. Wang, E. Yoon, N.-M. Hwang, D.-Y. Kim, K.M. Ho, Diffusion, coalescence, and reconstruction of vacancy defects in graphene layers. Phys. Rev. Lett. **95**, 205501 (2005)
49. O. Cretu, A.V. Krasheninnikov, J.A. Rodríguez-Manzo, L. Sun, R.M. Nieminen, F. Banhart, Migration and localization of metal atoms on strained graphene. Phys. Rev. Lett. **105**, 196102 (2010)
50. J. Kotakoski, A.V. Krasheninnikov, U. Kaiser, J.C. Meyer, From point defects in graphene to two-dimensional amorphous carbon. Phys. Rev. Lett. **106**, 105505 (2011)
51. J. Kotakoski, J.C. Meyer, S. Kurasch, D. Santos-Cottin, U. Kaiser, A.V. Krasheninnikov, Stone-wales-type transformations in carbon nanostructures driven by electron irradiation. Phys. Rev. B **83**, 245420 (2011)
52. L. Wang, X. Zhang, H.L.W. Chan, F. Yan, F. Ding, Formation and healing of vacancies in graphene chemical vapor deposition (CVD) growth. J. Am. Chem. Soc. **135**(11), 4476–4482 (2013)
53. S. Standop, O. Lehtinen, C. Herbig, G. Lewes-Malandrakis, F. Craes, J. Kotakoski, T. Michely, A.V. Krasheninnikov, Carsten Busse, Ion impacts on graphene/ir(111): interface channeling, vacancy funnels, and a nanomesh. Nano Lett. **13**(5), 1948–1955 (2013)

Chapter 2
Theoretical Modelling Methods

Density functional theory methods allow for the calculation of the total energy of a system of atoms and the forces acting on the individual atoms. Finding these forces allows the relaxation of structures to their lowest energy structure by driving the atomic forces towards zero. Calculation of the forces on a system also allows further quantities to be studied. The nudged elastic band (NEB) method can be used to find the minimum energy pathway across the potential energy surface between two minima by calculating the relevant forces along the pathway. This allows the saddle point and the corresponding energy barrier between two minima to be found, which can be used to determine the kinetics of various processes. Forces can also be used to determine the lattice dynamics of a system which can be used to study phonons and calculate the vibrational free energy. In molecular dynamics simulations the forces on each atom are evaluated at each sequential time step in order to evaluate the atomic motion.

A variety of these calculations are used in this thesis to study different properties involved in epitaxial graphene growth. In this chapter the methodology of density functional theory calculations as well as the DFT-based methods will be discussed in detail.

2.1 Density Functional Theory

2.1.1 Formalism

In density functional theory (DFT) the many-body Schrödinger equation for a system of electrons and nuclei interacting with each other is solved self-consistently. Using the Born-Oppenheimer approximation the nuclei are treated as being effectively fixed in position while the electrons move in the effective potential generated by them [1].

H.A. Tetlow, *Theoretical Modeling of Epitaxial Graphene Growth on the Ir(111) Surface*, Springer Theses, DOI 10.1007/978-3-319-65972-5_2

The Hohenberg-Kohn theorem states that the energy of a system of elections in the potential of the nuclei can be written as a functional of the electronic density $\rho(\mathbf{r})$ [2]:

$$E[\rho(\mathbf{r})] = F[\rho(\mathbf{r})] + \int d\mathbf{r} V_{ext}(\mathbf{r})\rho(\mathbf{r}) \tag{2.1}$$

where V_{ext} is the effective external potential of the nuclei and $F[\rho(\mathbf{r})]$ is an unknown universal functional of the electronic density that represents all the electron effects, which does not depend on the system in question. This will include their kinetic energy and the electron–electron interaction, $F[\rho(\mathbf{r})] = T[\rho(\mathbf{r})] + V_{ee}[\rho(\mathbf{r})]$, including all exchange-correlation effects. When $E[\rho(\mathbf{r})]$ is minimised with respect to the electronic density it corresponds to the ground state energy of the system E_0. Furthermore the exact electronic density which minimises $E[\rho(\mathbf{r})]$ is the ground state density $\rho_0(\mathbf{r})$. The ground state energy is given by

$$E_0 = E[\rho_0(\mathbf{r})] \leq E[\rho(\mathbf{r})] \tag{2.2}$$

In order to find the ground state energy the Kohn–Sham method is used [3]. To perform the minimisation the energy term $E[\rho(\mathbf{r})]$ is separated into its constituents which each have specific functional dependence on the electronic density $\rho(\mathbf{r})$. This maps the system to that of a system of non-interacting electrons within an effective potential. The electron density is described in terms of the effective one-particle wavefunctions $\psi_i(\mathbf{r})$ for a system containing N_e electrons:

$$\rho(\mathbf{r}) = \sum_{i=1}^{N_e} |\psi_i(\mathbf{r})|^2 \tag{2.3}$$

For the many-particle system the total energy is formed of the kinetic energy of the non-interacting electrons $T_{KS}[\rho(\mathbf{r})]$ and the potential energy of interactions between electrons $V_{KS}^{ee}[\rho]$ (the Hartree energy), as well as the external potential $\int d\mathbf{r} V_{ext}(\mathbf{r})\rho(\mathbf{r})$ due to the atomic cores:

$$E[\rho(\mathbf{r})] = T_{KS}[\rho(\mathbf{r})] + V_{KS}^{ee}[\rho(\mathbf{r})] + \int d\mathbf{r} V_{ext}(\mathbf{r})\rho(\mathbf{r}) + E_{xc}[\rho(\mathbf{r})]. \tag{2.4}$$

The form of the Hartree energy is $V_{KS}^{ee}[\rho] = \frac{e}{2}\int \frac{\rho(r)\rho(r')}{|r-r'|} dr dr'$. Errors in the kinetic energy due to interactions are incorporated into the exchange-correlation term $E_{xc}[\rho(\mathbf{r})]$, as well as the corrections to the electron–electron interaction. The full form of $E_{xc}[\rho(\mathbf{r})]$ is therefore

$$\begin{aligned} E_{xc}[\rho(\mathbf{r})] &= F[\rho(\mathbf{r})] - V_{KS}^{ee}[\rho(\mathbf{r})] - T_{KS}[\rho(\mathbf{r})] \\ &= (V^{ee}[\rho(\mathbf{r})] - V_{KS}^{ee}[\rho(\mathbf{r})]) + (T[\rho(\mathbf{r})] - T_{KS}[\rho(\mathbf{r})]) \end{aligned} \tag{2.5}$$

Developing the correct form of $E_{xc}[\rho(\mathbf{r})]$ is crucial for generating a more accurate functional. This will be discussed in Sect. 2.1.2.

Minimisation of the energy functional (2.4) allows the formulation of an effective potential. The ground state density of this many particle system is then equivalent to that of a system of non-interacting electrons moving in this effective external potential field given by

$$V_{eff}(\mathbf{r}) = V_{ext}(\mathbf{r}) + \int d\mathbf{r} \frac{\rho(\mathbf{r}')}{|\mathbf{r} - \mathbf{r}'|} + \frac{\delta E_{xc}}{\delta \rho(\mathbf{r})}. \tag{2.6}$$

The Kohn–Sham Hamiltonian is formed of $V_{eff}(\mathbf{r})$ and the kinetic energy of the "fictitious" non-interacting electron gas [3],

$$H_{KS} = -\frac{1}{2}\nabla^2 + V_{eff}(\mathbf{r}). \tag{2.7}$$

This Hamiltonian has no direct interactions between electrons, but includes the effect of each electron moving in a potential field dependent on the distribution of electrons $\rho(\mathbf{r})$. This solves the Schrödinger equation,

$$\left[-\frac{1}{2}\nabla^2 + V_{eff}(\mathbf{r})\right] \psi_i(\mathbf{r}) = \epsilon_i \psi_i(\mathbf{r}). \tag{2.8}$$

Solving the Kohn–Sham equations [3] to find the ground state density is achieved self-consistently by an iterative process. First a starting density $\rho(\mathbf{r})$ is used to calculate the effective potential and solve the Schrödinger equation. From this a new density is constructed which is used to generate new one-electron wavefunctions and restart the iteration. This process ends when the density has converged to within a certain error. This value will correspond to the ground state.

2.1.2 The Exchange-Correlation Functional

The main cause for inaccuracy in this Kohn–Sham method is in the approximation needed for the exchange and correlation effects. Choosing the correct form of $E_{xc}[\rho]$ is important in order to reduce the error. Popular choices for this are the local-density approximation (LDA) and the generalised gradient approximation (GGA).

In the LDA the exchange-correlation functional is expressed only in terms of the electron density,

$$E_{xc} = \int \rho(\mathbf{r})\epsilon_{xc}(\rho)d\mathbf{r} \tag{2.9}$$

where ϵ_{xc} is the exchange-correlation energy per particle [4]. In Eq. (2.9) it is assumed that the electron density changes smoothly throughout space. In the GGA E_{xc} is expressed in terms of the electron density and the electron density gradient $\nabla\rho(\mathbf{r})$,

$$E_{xc} = \int \rho(\mathbf{r})\epsilon_{xc}(\rho, \nabla\rho)d\mathbf{r}. \tag{2.10}$$

This improves the exchange-correlation energy in situations when the density varies rapidly [5, 6]. Both the LDA and GGA functionals do not take into account any non-local interactions, such as the dispersion interaction. To include van der Waals (dispersion) forces, additional functionals are needed. These are discussed in Sect. 2.1.3.

2.1.3 Van der Waals Forces in DFT

An important consideration in density functional theory calculations is the ability to incorporate van der Waals (vdWs) forces. These forces play a major role in many systems (such as molecules and proteins) and affect their structure and stability. They arise from the dispersion interaction. Charge fluctuations in different parts of an atomic system become correlated over long ranges which results in the attractive vdWs force. However DFT based on standard functionals such as LDA and GGA fails to correctly model vdWs forces.

In order to model dispersion forces many DFT based techniques have been developed with varying accuracy. Some methods (semi-empirical) rely on input parameters to formulate a description of the dispersion energy whereas others calculate it directly from the electron density. Some of these various techniques are described below.

2.1.3.1 DFT-D

The DFT-D method includes the dispersion energy by adding an extra density-independent term to the energy density functional. With this formalism the energy becomes:

$$E_{tot} = E_{DFT} + E_{disp}. \tag{2.11}$$

The E_{disp} term accounts for the long range interaction between a pair of atoms which decays as $-1/r^6$ with the atomic separation r. The full expression for E_{disp} is:

$$E_{disp} = -\sum_{A,B} C_6^{AB}/r_{AB}^6 \tag{2.12}$$

which includes the dispersion coefficients C_6^{AB} that depend on the nature of atoms A and B. This interaction is isotropic and these coefficients are taken as being constant during the calculation, hence they do not depend on the environment of the atoms. Obtaining a good universal choice for them is difficult as it relies on experimental parameters which may not be available for some elements. Other problems with this method include the neglect of many body dispersion effects and higher order (C_8/r^8, C_{10}/r^{10}) terms. Also for small r the energy diverges.

2.1.3.2 DFT-D2

The DFT-D2 method proposed by Grimme in 2006 offers an improvement to the DFT-D method [7]. In this formalism the dispersion energy contribution is modified (see below) and the C_6 coefficients are calculated from the product of the ionisation potentials and the static polarisabilities of isolated atoms. This allows coefficients for a greater number of elements to be included. However in some cases the coefficients can not be correctly derived from this expression and instead averages from other elements are used.

In order to prevent the divergence of E_{disp} which occurs at small r a damping function $f(r_{AB}, A, B)$ is added into the DFT-D2 E_{disp} expression:

$$E_{disp} = -\sum_{A,B} f(r_{AB}, A, B) C_6^{AB} / r_{AB}^6. \tag{2.13}$$

One choice of the damping function is the Fermi function,

$$f(r_{AB}, A, B) = \frac{1}{1 + e^{-d(r_{AB}/R_r - 1)}} \tag{2.14}$$

where r_{AB} is the distance between two atoms A and B, R_r is the sum of atomic vdW radii and d is the steepness of the damping function. The choice of the damping function is important and must be constructed so that it fits with reference data. It can be chosen so that it includes the effect of higher order (C_8/r^8, C_{10}/r^{10}) terms [8]. Its form will also be dependent on the type of exchange-correlation functional to be used alongside this correction.

2.1.3.3 DFT-D3 and vdW(TS)

A problem with the DFT-D2 method is that the C_6 coefficients still have no dependence on the environment of the atom. The DFT-D3 and vdW(TS) methods improve on this by allowing the dispersion coefficient to depend on the effective volume of the atom. If an atom is effectively squeezed by other atoms then the C_6 coefficient is made smaller. In DFT-D3 the number of neighbours that a particular atom has is calculated and used to adjust the coefficient [8]. For a large number of neighbours the effective volume is assumed to be smaller and therefore C_6 is reduced. In this way the coefficients can be adjusted throughout the calculation. In the vdW(TS) method (proposed by Tkatchenko and Scheffler) effective atomic volumes are used [9]. The total electron density of a molecule is split between its individual atoms and this is then compared to the density of a free atom. The C_6 coefficient is scaled depending on the difference.

2.1.3.4 vdW-DF

The van der Waals density functional (vdW-DF) method calculates the dispersion from the electron density [10, 11]. For this the exchange correlation energy functional is divided into the original exchange term and a new correlation term,

$$E_{xc}[\rho(\mathbf{r})] = E_x^{GGA}[\rho(\mathbf{r})] + E_c^{new}[\rho(\mathbf{r})]. \tag{2.15}$$

$E_x^{GGA}[\rho(\mathbf{r})]$ is the exchange functional as determined by a GGA technique and $E_c^{new}[\rho(\mathbf{r})]$ is a non-local correlation function which includes vdWs interactions. This in turn can be divided into two different contributions:

$$E_c^{new}[\rho(\mathbf{r})] = E_c^0[\rho(\mathbf{r})] + E_c^{nl}[\rho(\mathbf{r})]. \tag{2.16}$$

$E_c^0[\rho(\mathbf{r})]$ can be approximated with the LDA correlation function $E_c^0[\rho(\mathbf{r})] \approx E_c^{LDA}[\rho(\mathbf{r})]$. $E_c^{nl}[\rho(\mathbf{r})]$ represents the long ranged non-local interactions which give rise to vdWs. It has the form of:

$$E_c^{nl}[\rho(\mathbf{r})] = \frac{1}{2}\int d^3\mathbf{r} \int d^3\mathbf{r}' \rho(\mathbf{r})\Phi(\mathbf{r}, \mathbf{r}')\rho(\mathbf{r}') \tag{2.17}$$

where $\Phi(\mathbf{r}, \mathbf{r}')$ is an integration kernel. A problem with vdW-DF is that is tends to overestimate the long ranged interactions. The choice of the $E_x^{GGA}[\rho(\mathbf{r})]$ functional can also have an effect on the accuracy of the result. The vdW-DF method is more expensive than the Grimme-like methods. Furthermore a "universal" vdW functional does not currently exist [12].

2.2 Basis-Sets

Before solving the Kohn–Sham equations first the form of the wavefunctions, $\psi_i(\mathbf{r})$ must be chosen. The wavefunction is expanded into a set of basis functions, ϕ_j with coefficients c_{ij}

$$\psi_i = \sum_j c_{ij}\phi_j. \tag{2.18}$$

The type of the basis set is important as it affects how computationally expensive it is to converge the system with the required accuracy. The choice of the basis set is also dependent on the type of system. For a crystalline solid the basis set will obey Bloch's theorem since the system is periodic. Due to the periodicity the potential $V(\mathbf{r})$ will have the form

$$V(\mathbf{r}) = V(\mathbf{r} + \mathbf{R}) \tag{2.19}$$

where $\mathbf{R} = n_1\mathbf{a}_1 + n_2\mathbf{a}_2 + n_3\mathbf{a}_3$ is the direct lattice vector, with $\mathbf{a}_{1,2,3}$ the unit cell vectors of the system. Based on this the basis functions $\phi_{\mathbf{k}}(\mathbf{r}) = u_{\mathbf{k}} e^{i\mathbf{k}\cdot\mathbf{r}}$ will satisfy the relation

$$\phi_{\mathbf{k}}(\mathbf{r} + \mathbf{R}) = e^{i\mathbf{k}\cdot\mathbf{R}} \phi_{\mathbf{k}}(\mathbf{r}) \tag{2.20}$$

where $\mathbf{k}$ is the reciprocal space wave vector and $u_{\mathbf{k}}$ are arbitrary functions that obey the periodicity of the lattice. A set of plane waves fulfil this condition and are commonly used for a system with periodic boundary conditions [13]. Then the one-electron wavefunctions ψ_i can be expanded in them:

$$\psi_{i,\mathbf{k}}(\mathbf{r}) = \sum_{\mathbf{G}} c_{i,\mathbf{k}}(\mathbf{G}) e^{i(\mathbf{G}+\mathbf{k})\cdot\mathbf{r}} \tag{2.21}$$

Here $\mathbf{G} = m_1\mathbf{b}_1 + m_2\mathbf{b}_2 + m_3\mathbf{b}_3$ is the reciprocal lattice vector and $c_{i\mathbf{k}}(\mathbf{G})$ are expansion coefficients. The advantages of using plane waves is that they are orthogonal and Fast Fourier Transforms can be used in the calculation of the coefficients $c_{i,\mathbf{k}}(\mathbf{G})$. However when using a plane wave basis set for large systems many plane waves are needed. Also the wavefunctions oscillate within the core region and as a result the use of pseudopotentials is required (see Sect. 2.3).

Another choice of basis set is a linear combination of atomic orbitals (LCAO). Atomic orbitals decay to zero at infinite distances in the same way as the wavefunction of an isolated atom or molecule. Therefore these are a good choice of basis function. They have the form:

$$\phi_{\mathbf{j}}(\mathbf{r}) \rightarrow \chi_{nlm}(\mathbf{r}) = R_{nl}(r) S_{lm}(\mathbf{r}) \tag{2.22}$$

where $S_{lm}(\mathbf{r})$ is a spherical harmonic function [13], $R_{nl}(r)$ is the radial function and n, l, m are the usual quantum numbers for atomic orbitals. A minimum basis set is constructed, formed of the minimum necessary number of atomic orbitals for the system. However this minimum basis set fails at describing the chemical bonding and converging the energy of the system. To account for this diffuse atomic orbitals can be added to increase the basis set expansion and give a better representation of the charge density.

A further problem of LCAOs is the basis set superposition error (BSSE). This occurs when two species with different basis sets approach one another and one species may utilise the basis sets of the other species (or vice versa) to describe its electron distribution. This will improve the description of both species when they are interacting. However this description can now not be compared with when they are non interacting at large separations and hence the treatment is inconsistent. The BSSE can be corrected for by calculation of the counterpoise correction [14]. Furthermore there is a so-called Pulay correction to the atomic forces when LCAOs are used [13] due to the fact that the positions of the AOs depend explicitly on atomic positions.

The radial part of the atomic orbital may have the form of a Slater type orbital, $R_{nl}(r) \sim e^{-\xi r}$ or a Gaussian type orbital $R_{nl}(r) \sim e^{-\alpha r^2}$. Despite the fact that many Gaussian type orbitals are needed in order to get a good representation, they are computationally less expensive than Slater type orbitals and are therefore the favourable choice. Compared to plane waves, less Gaussian type orbitals are required in order to represent the wavefunction. To represent the basis set, minimal (one basis function per AO), double zeta (two basis functions per AO), triple zeta (three basis functions per AO) or a split valance (one basis function per core AO, and more for the valence AOs) basis sets may be used.

Different electronic structure codes use different choices for basis sets. For example, the Vienna *ab initio* simulation package (VASP) [15, 16] uses a plane wave basis set whereas the CP2K code relies on both Gaussians and plane waves (GPW) [17]. The wavefunction is described using a combination of Gaussian type orbitals, but the electron density is expanded in plane waves to facilitate the calculation of the Coulomb potential; the exchange-correlation energy then is calculated on a grid. For large systems this method scales as $O(N_M^2 M)$ where M is the number of basis functions and N_M is the number of molecular orbitals.

2.2.1 K-Point Sampling

So far the quantities discussed have been considered for an infinite k-space sampling in the first Brillouin zone. However in practice the size of the k-mesh is limited. In order to integrate a function over the Brillouin zone, k-point sampling must be used. For example the electron density is given by

$$\rho(\mathbf{r}) = \frac{1}{\Omega_{BZ}} \sum_{n}^{occ} \int_{BZ} |\psi_n(\mathbf{r})|^2 d^3\mathbf{k} \tag{2.23}$$

where the sum is performed over all occupied states in the Brillouin zone and Ω_{BZ} is the volume of the Brillouin zone. Only a finite number of k-points are possible, and the number required depends on the system in question. Usually k-points are determined using the Monkhorst-Pack method [13, 18].

VASP allows the use of a k-mesh for integrating over in reciprocal space (k-space) to calculate properties. For larger cells a smaller k-mesh is required, and inversely small cells need a greater number of k-points. Metals in particular need more k points as the interaction between the conduction electrons are long ranged. In VASP small real space (r-space) cells may be used, as the k-points can compensate for the size of the cell. As the computation time scales linearly with number of k-points but to the order of N_A^3 for number of atoms this is a much cheaper alternative. By examining the convergence of total energy with k-points the size of the k-mesh can be determined. In CP2K only the gamma point is used, therefore there is not a well defined k-mesh for the integration. To compensate for this large r-space cells must be used so that the k-space sampling size is reduced.

2.3 Pseudopotentials

The electrons in atoms can be considered in terms of two groups: inner core electrons and valence electrons. The inner core electrons are tightly bound to the nucleus and play a limited part in the chemical bonding with other atoms. The valence electrons are screened from the effects of the nucleus by the core electrons and are involved in bonding. During bonding the wavefunctions of the core electrons are only slightly affected as they remain strongly localised around their cores and therefore they can be considered as essentially inert. These electrons along with the nucleus can be treated with non-variational wavefunctions to simplify the method. This is known as the frozen core approximation.

In addition to this, a problem arises due to the fact that the valence electrons are required to be orthogonal to the core electrons within the core region. This causes the spin orbitals of the valence electrons to oscillate in the core region and as a result of these oscillations many basis functions are needed in order to describe them correctly. Therefore, when using a plane wave basis set, a large value of the cutoff radius, G_{max} is required (see Eq. (2.21)), with a high energy cutoff E_{cut} for the wavefunctions $E_{cut} = \hbar^2 G_{max}^2 / 2m$. This increases the computational effort.

To reduce the number of basis sets needed, pseudopotentials are used to describe the Coulomb potential of the valence electrons as shown in Fig. 2.1. These are modified to ensure that the pseudo-wavefunctions are the same as the actual wavefunctions outside the core radius and smooth inside the core radius.

Common forms of the pseudopotentials include the norm-conserving, ultrasoft and the projected-augmented wave method (PAW) [13]. The PAW method (used in VASP) allows an effectively all-electron calculation with frozen core orbitals (the other methods are based on valence pseudo-wavefunctions) [13]. It works by splitting the valence electron function ϕ_v into three parts, such that $\phi_v = \tilde{\phi}_v + \chi_v + \tilde{\chi}_v$. $\tilde{\phi}_v$ is smooth everywhere and equal to ϕ_v outside the core (augmented) region, whereas χ_v is equal to the true orbital ϕ_v in the core region and smooth outside it. The net differences are included in $\tilde{\chi}_v$. $\tilde{\phi}_v$ is used to represent the pseudo-wavefunction.

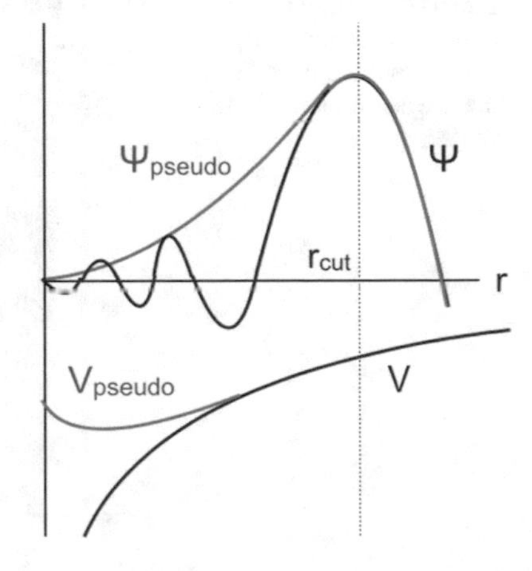

Fig. 2.1 The pseudopotentials V$_{pseudo}$ replaces the Coulomb potential V and as a result the oscillating core wavefunctions ψ are replaced by the pseudo-wavefunctions ψ_{pseudo} which are smooth inside the core region

The Goedecker-Teter-Hutter (GTH) [19] method is used in the CP2K calculations of this thesis. Here the pseudopotentials are separated into local and non-local parts. The pseudopotentials are represented in an analytical form, with the parameters required to characterise each element in the form of Gaussians functions. This improves the efficiency.

2.4 The Nudged Elastic Band Method

The nudged elastic band method (NEB) is a valuable technique that finds the minimum energy pathway (MEP) between two minima on the potential energy surface [20–22]. This means that details of a transition from one state to another can be determined, such as the energy of the saddle point and its structure. This energy corresponds to the maximum energy point along the MEP and dictates the energy barrier that needs to be overcome during a transition. This technique is particularly useful for determining the rate of a processes within transition state theory using the Arrhenius equation (see Sect. 2.7.1).

During a NEB calculation an elastic band connecting the two minima is optimised to find the MEP. The optimisation acts as to minimise the force perpendicular to the band and the force along the band, which is treated as elastic with a spring constant k_{spring}. To begin, a number of images are constructed between the initial and final states which act as an initial guess for the path. For a simple case, such as the diffusion of an atom on a surface from one site to another, the images may be generated by linear interpolation (Fig. 2.2).

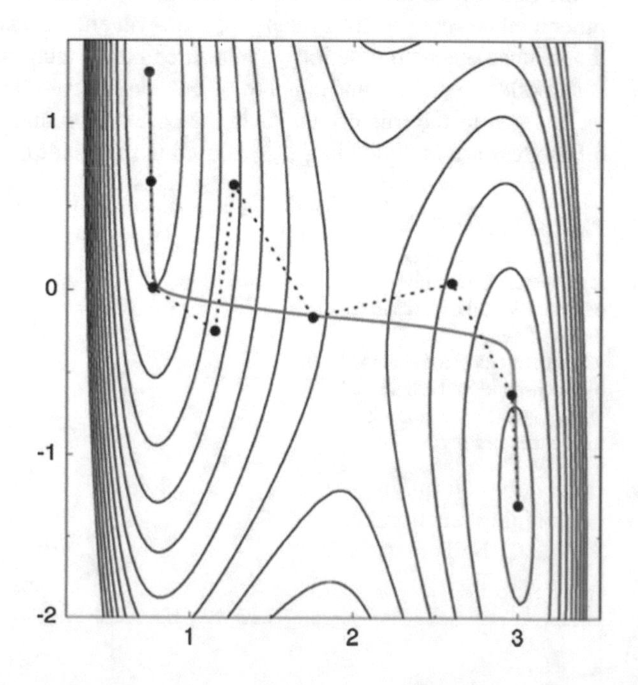

Fig. 2.2 The potential energy surface showing two minima connected by the initial energy pathway generated from interpolation (*dotted line*) and the optimised minimum energy pathway as calculated by NEB. Reprinted from [21], with the permission of AIP Publishing

During the NEB optimisation process the position of the images are adjusted. For $N+1$ images, the position of each freely moving atom corresponds to a vector $[\mathbf{R}_0, \mathbf{R}_1, \ldots, \mathbf{R}_N]$ (with fixed endpoints $\mathbf{R}_0$ and $\mathbf{R}_N$). The force on each image is calculated including the component of the spring force parallel to the tangent of each image and the component of the true force perpendicular to the tangent. The tangent for each image is estimated from [21],

$$\hat{\tau}_i = \frac{\mathbf{R}_{i+1} - \mathbf{R}_{i-1}}{|\mathbf{R}_{i+1} - \mathbf{R}_{i-1}|}. \tag{2.24}$$

The total force is then,

$$\mathbf{F}_i = \mathbf{F}_i^s|_{\|} - \nabla V(\mathbf{R}_i)|_{\perp} \tag{2.25}$$

which is comprised of the perpendicular true force,

$$\nabla V(\mathbf{R}_i)|_{\perp} = \nabla V(\mathbf{R}_i) - \nabla V(\mathbf{R}_i) \cdot \hat{\tau}_i \tag{2.26}$$

and the spring force,

$$\mathbf{F}^s{}_i|_{\|} = k_{spring}[(\mathbf{R}_{i+1} - \mathbf{R}_i) - (\mathbf{R}_i - \mathbf{R}_{i-1})] \cdot \hat{\tau}_i \hat{\tau}_i. \tag{2.27}$$

The total force calculated for each image is applied to move their atomic positions at each step of the optimisation process and a projected velocity Verlet algorithm is used for the minimisation. The process continues until the force on the images has converged to zero.

For NEB it is advised to use an odd number of images as one image (most likely in the middle) has a greater chance of finding the transition state, and will therefore give a more accurate value for the saddle point energy. Ideally many images should be used, as this will be able to give a better description of the minimum energy pathway. However adding images is computationally expensive since the number of processors required scales linearly with the number of images.

2.4.1 Climbing Image NEB

The climbing image method (CI-NEB) [20] improves the finding of the saddle point. The image with the highest energy is made to climb up the minimum energy pathway towards the saddle point and the spring constant across the band is varied so that the density of images is greatest near the saddle point. This is particularly useful when accurately determining the energy of the saddle point is required, such as when determining reaction rates (Fig. 2.3).

The CI-NEB method works by identifying the highest energy image after a few iterations. Then the force on this image is modified to,

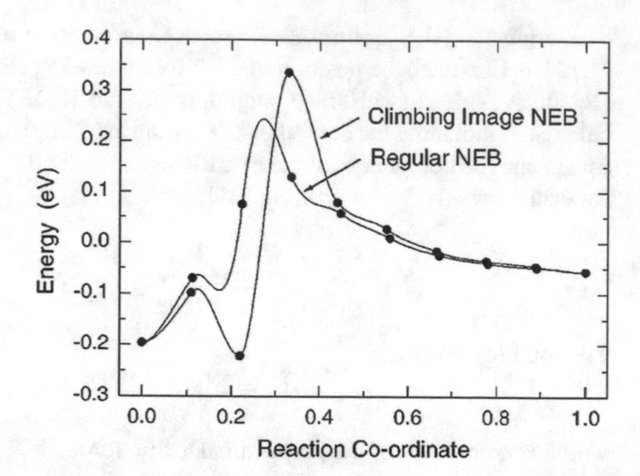

Fig. 2.3 The energy profile of the elastic band calculated with CI-NEB compared to regular NEB. The climbing image converges to find the true saddle point. Reprinted from [20], with the permission of AIP Publishing

$$\mathbf{F}_{i_{max}} = -\nabla V(\mathbf{R}_{i_{max}}) + 2(\nabla V(\mathbf{R}_{i_{max}}) \cdot \hat{\tau}_{i_{max}})\hat{\tau}_{i_{max}}. \tag{2.28}$$

So that the force is inverted and the spring force component is completely removed. This force will now drive this image to eventually converge more closely to the saddle point.

2.5 Lattice Dynamics

Calculating the lattice dynamics of a system using DFT is a useful tool. It can be used to find the vibrational modes (phonons), and in turn find the phonon vibrational frequencies ω and the corresponding phonon density of states $D(\omega)$. The vibrational frequencies of a system are necessary for calculating the vibrational component of the free energy of a system. Furthermore, in transition state theory (Sect. 2.7.1) the pre-exponential factors A in an Arrhenius equation for the rate of a process $R = A\exp(-E/k_BT)$, with an energy barrier E, can be determined by finding the frequencies at the initial state and the transition state using the Vineyard equation [23] (see Eq. (2.43)). Finally, calculating the frequencies is needed in order to confirm convergence to the true saddle point since at this point there must be exactly one imaginary frequency.

In order to determine the lattice dynamics the dynamical matrix may be calculated using the frozen-phonon approximation [24]. The starting point for this is the Hamiltonian for the motion of atoms in a crystal. This is comprised of the kinetic energy and the potential energy,

$$H = \sum_{\alpha,k} \frac{p_{\alpha}^2, k}{2m_k} + \Phi \tag{2.29}$$

The potential energy Φ is the result of small atomic displacements. For a system of atoms in a unit cell with periodic boundary conditions applied, the displacement of an atom, κ, in a unit cell, l is given by, $\mathbf{u}_{l\kappa}(t)$. Therefore at a time t the position of a particular atom κl is given by $\mathbf{R}_{kl} = \mathbf{r}_k + \mathbf{r}_l + \mathbf{u}_{\kappa l}(t)$.

The potential energy can be Taylor expanded in terms of the atomic displacements about the equilibrium positions of the atoms:

$$\Phi = \Phi_0 - \sum_{\alpha' l' k'} \Phi_{\alpha' l' k'} u_{\alpha' l' k'} + \frac{1}{2} \sum_{\alpha lk, \alpha' l' k'} \Phi_{\alpha lk, \alpha' l' k'} u_{\alpha lk} u_{\alpha' l' k'} + \cdots \tag{2.30}$$

where the displacement direction, α is one of the cartesian components x, y, z. The first order term,

$$\left(\frac{\partial \Phi}{\partial u_{\alpha lk}} \right)_0 = -\Phi_{\alpha' l' k'} \tag{2.31}$$

becomes zero, since at equilibrium the forces are zero. Hence the potential (in the harmonic approximation) becomes

$$\Phi = \Phi_0 + \frac{1}{2} \sum_{\alpha lk, \alpha' l' k'} \Phi_{\alpha lk, \alpha' l' k'} u_{\alpha lk} u_{\alpha' l' k'} \tag{2.32}$$

where $\Phi_{\alpha lk, \alpha' l' k'}$ is the force constant matrix

$$\Phi_{\alpha lk, \alpha' l' k'} = \left(\frac{\partial^2 \Phi}{\partial u_{\alpha lk} u_{\alpha' l' k'}} \right)_0 \tag{2.33}$$

which describes the force on an atom kl in the direction α due to the motion of an atom $k'l'$ displaced in the direction α'. Based on the force constant matrix and its symmetry, $\Phi_{\alpha lk, \alpha' l' k'} = \Phi_{\alpha' l' k', \alpha lk}$, the equation of motion for the system is

$$M_k \frac{\partial^2 u_{\alpha lk}}{\partial t^2} = - \sum_{\alpha' l' k'} \Phi_{\alpha lk, \alpha' l' k'} u_{\alpha' l' k'} \tag{2.34}$$

with M_k the atomic mass. This can be solved by seeking solutions of the form, $u_{\alpha lk} = 1/\sqrt{M_k} c_{\alpha lk} \exp(-i\omega t) \exp(-ikR)$, to give,

$$\omega^2 c_{\alpha lk} = \sum_{\alpha' l' k'} \frac{\Phi_{\alpha lk, \alpha' l' k'}}{\sqrt{M_k M_k'}} c_{\alpha' l' k'} \tag{2.35}$$

This can be expressed in matrix form as an eigenvalue and eigenvector problem:

$$\omega^2 \mathbf{c_k} = \mathbf{D_k} \mathbf{c_k} \tag{2.36}$$

with $\mathbf{c_k} = (c_{\alpha lk})$ a matrix of amplitudes, and $\mathbf{D_k}$ the dynamical matrix with elements:

$$D_{\alpha lk,\alpha' l' k'}(k) = \frac{1}{\sqrt{M_k M_k'}} \sum_{\alpha' l' k'} \Phi_{\alpha lk,\alpha' l' k'} \exp(-ikR). \tag{2.37}$$

Determining the dynamical matrix is achieved by the calculation of the force constant matrix. This contains $(3N)^2$ elements due to the cartesian components for each atom. By displacing each atom sequentially in the different directions and then calculating the resultant forces on the atoms, the individual elements can be calculated. For example, for an atom a displaced along the x-direction the force on atom b in the y direction gives rise to the force constant element $\Phi_{ax,by}$. This force constant component is given by,

$$\Phi_{ax,by} = \frac{F_{b,y}(a,+x) - F_{b,y}(a,-x)}{2\Delta} \tag{2.38}$$

where $F_{b,y}(a,+x)$ and $F_{b,y}(a,-x)$ are the forces on atom b in the y-direction, due to the displacement of atom a in the positive and negative x-directions respectively, and Δ is the length of the displacement. The value of Δ must be small with respect to the inter-atomic distance.

Once the force constants are found the eigenvalue problem Eq. (2.36) is solved to find the 3 N eigenvalues ω^2, which are the vibrational frequency modes. These may be used to determine the Arrhenius pre-factors in transition state theory as described in Sect. 2.7.1, or to calculate the vibrational free energy.

2.5.1 *Vibrational Free Energy*

The vibrational free energy due to the phonons in a system can be calculated using the harmonic approximation. The partition function for a harmonic solid with N atoms is simply the product of multiple single harmonic oscillators,

$$Z_N = \prod_{i=1}^{3N} \frac{e^{-\beta\epsilon_i/2}}{1 - e^{-\beta\epsilon_i}}. \tag{2.39}$$

where $\beta = 1/k_b T$ and $\epsilon_i = \hbar\omega_i$ is the energy of an individual phonon with frequency ω_i. The free energy $F_{vib} = -k_B T \ln(Z_N)$, is then given by,

$$F_{vib} = \frac{1}{2}\sum_{i=1}^{3N} \epsilon_i + k_B T \sum_{i=1}^{3N} \ln(1 - e^{-\beta\epsilon_i}). \tag{2.40}$$

Once the vibrational frequency modes are known, the vibrational free energy can then be calculated.

2.6 Core Level Binding Energies

X-ray photoemission spectroscopy experiments are performed by measuring the binding energy of a core electron. This electron is excited from the core by the x-ray photon and its kinetic energy left after leaving the surface (i.e. spent on the work function) is measured to determine its binding energy [25].

The core level binding energy of an electron in a particular atom can be calculated approximately with DFT methods. In this method the electron is removed from the core to the valence band. To calculate the binding energy the energy difference between a system with all electrons in the unexcited ground state, and the system with the electron excited from the core is determined. This is given by the energy difference,

$$E_{CLS} = E(n_c - 1) - E(n_c) \tag{2.41}$$

where $E(n_c)$ is the energy of the ground state with core electrons n_c, and $E(n_c - 1)$ is the energy of the excited state with $n_c - 1$ core electrons. The excited state is considered by removing the electron from the core and then relaxing the remaining core electrons, during which the core ionic PAW potential is created. The calculation is approximate as the core electron is not completely removed from the system. This would result in the system being charged, which is undesirable in calculations with periodic boundary conditions.

Calculating the core binding energy is useful as it allows experimental XPS spectra to be analysed. Each atom in a different environment will have a different core level binding energy. Based on this different species can be identified. It is worth noting that the calculated core binding energies can not be compared directly to the experimental binding energies and some relative shift will need to be applied to the energies in order to compare the data.

2.7 Kinetics

In order to the determine the kinetics of various important chemical processes it is necessary to be able to evolve the dynamical motion of an atomic system in time. A simple way to do this is with molecular dynamics (MD) simulations. The forces on each atom are determined (classically or with DFT) and then the Newtonian equations of motion are solved to evolve the system (see Sect. 2.8). However this becomes problematic when the particular processes of interest occur infrequently. In order to resolve the atomic vibrations the MD time step must be of the order of femtoseconds but these infrequent events may only occur on average every microsecond, thus requiring impractically long simulation times. A way to overcome this problem is to coarse-grain the frequent events and so only consider the transitions corresponding to infrequent events. This is the principle of kinetic Monte Carlo simulations, as illustrated in Fig. 2.4.

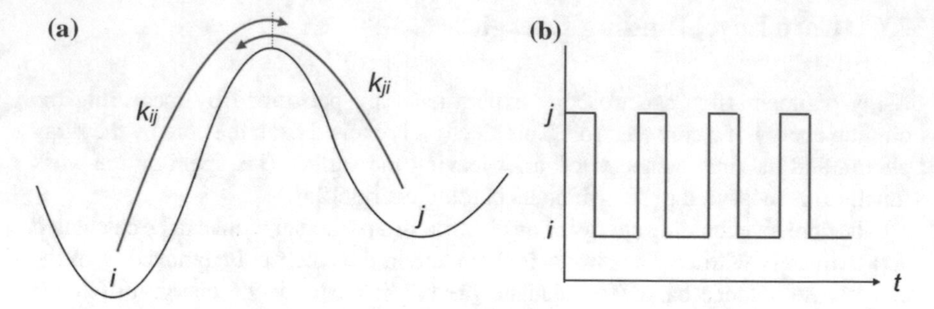

Fig. 2.4 An infrequent event as described by **a** molecular dynamics and **b** kinetic Monte Carlo. The event taking state i into state j has a low escape probability and hence the system will spend a long time in state i before the transition occurs. In molecular dynamics simulations this would result in long simulation times which is inefficient. In kinetic Monte Carlo the system is coarse-grained in terms of the atomic vibrations and only discrete events are considered

In order to coarse-grain and evolve only the infrequent events a Markov chain Master equation describing transitions out of a state is constructed. The rate of change in probability for being in a state i is found from the probability difference of entering the state from another state j, and leaving the state to another state j

$$\frac{dP_i}{dt} = -\sum_{j \neq i} k_{i \to j} P_i(t) + \sum_{j \neq i} k_{j \to i} P_j(t) \tag{2.42}$$

Here the rate constants $k_{i \to j}$ give the average escape rate from state i to another state j. The sum is taken over all states j. If the possible transitions and their rate constants are known then evolving the system is simply achieved by solving the set of equations (2.42). This can be achieved numerically using a kMC simulation, whereby an ensemble of trajectories is created, and then the system is evolved from state to state via the allowed trajectories. The probability for being in a state $P_i(t)$ can then be determined from averaging over many trajectories.

2.7.1 Transition State Theory

Transition state theory (TST) allows the calculation of the rate constant for a particular transition. For an infrequent transition between two states A and B (as shown in Fig. 2.5) the rate constant is given by the equilibrium flux through the dividing surface between the states. This approximation requires that the transitions are classical, and that there no subsequent re-crossings of the barrier. Transitions must be rare so that the system has enough time to equilibrate at the minimum and there should be no correlated dynamical events, i.e. it is assumed that the system spends long enough time vibrating in a basin that it has no memory of previous transitions. Once

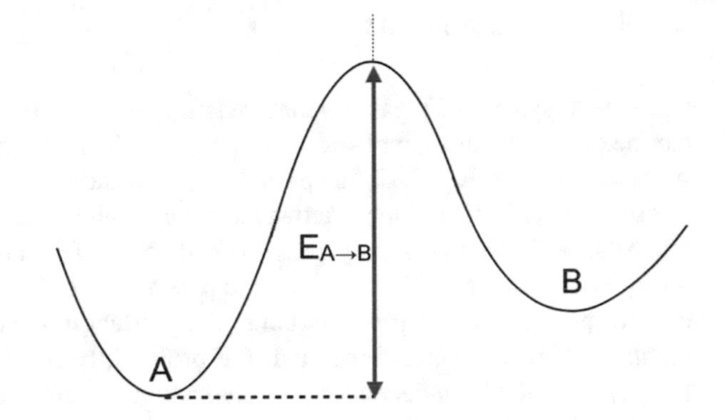

Fig. 2.5 For transition from A→B transition state theory can be used to determine the rate constant, $k_{A \rightarrow B}$ in terms of the energy barrier at the saddle point, $E_{A \rightarrow B}$ and the vibrational frequencies at the initial state and the saddle point (Eq. 2.43)

the system is at the dividing surface it is presumed that it will make the transition into the next state and not return to the original state. In fact this may lead to an overestimation in the rates [26, 27].

To further simplify transition state theory the harmonic approximation can be used (hTST) [23, 28]. The transition at the dividing surface is characterised by the saddle point which corresponds to the maximum along the transition pathway. Furthermore the potential energy at the minimum and at the saddle point are approximated by second-order potentials which correspond to the harmonic approximation. With this description the rate constant for the transitions from A to B can be given by

$$k^{hTST} = \frac{\prod_i^{3N} \nu_i^{min}}{\prod_i^{3N-1} \nu_i^{sad}} \exp(-E_{A \rightarrow B}/k_B T) \tag{2.43}$$

where $E_{A \rightarrow B}$ is the energy barrier given by the energy difference between the saddle point of the transition and the minimum of state A and k_B is the Boltzmann constant, T is the temperature. The pre-exponential factor is given by the ratio of the products of the $3N$ vibrational frequencies at the minimum and the $3N - 1$ frequencies at the saddle point i.e. with the imaginary frequency at the saddle point excluded, where N is the total number of atoms in the system. This ratio picks out the frequency at the minimum associated with the transition. Often the pre-exponential factor for a transition will be in the range of 10^{12} – 10^{13} s^{-1}.

Assuming both the initial and final states of a transition are known then the saddle point (and its energy barrier) can be found using the nudged elastic band method. Providing the saddle point is found accurately, the pre-exponential factors can also be determined by calculating the vibrational frequencies at the saddle point and either of the two minima. This allows the rate constant for a process to be found. In order to accurately simulate the kinetics of a system it is necessary to predetermine all possible processes and their energetics in this way. If certain transitions are overlooked then the correct reaction mechanisms may be excluded from the simulation, resulting in different outcomes to reality. This is one of the main obstacles for performing infrequent event simulations accurately.

2.7.2 Rate Equations

Consider a system of a gas of many reacting species. The evolution of this system can be calculated in a simplistic way using Eq. (2.42). In the limit of a large number of species then in Eq. (2.42) the probability of a state P_i can be approximated as the relative concentration of the reacting state constitutes N_i. For unimolecular reactions (A $\rightarrow$ B + C) this is simply proportional to the reactant concentration N_A. For bimolecular reactions (B + C $\rightarrow$ A) the probability of having the reacting state AB can be approximated as proportional to the product of the concentrations of the two reactants $N_A N_B$. This assumes that the probability for A and B to meet is equal to $N_A N_B$, which is independent of any spatial effects such as their diffusion. For instance, for a simple model system described by $A + B \leftrightarrow C$ with the rates of the forward and reverse processes given by $k_{AB,C}$ and $k_{C,AB}$ respectively, the Master equations describing the change in concentration for each species become:

$$\frac{dN_A}{dt} = N_C k_{C,AB} - N_A N_B k_{AB,C} \tag{2.44}$$

$$\frac{dN_B}{dt} = N_C k_{C,AB} - N_A N_B k_{AB,C} \tag{2.45}$$

$$\frac{dN_C}{dt} = N_A N_B k_{AB,C} - N_C k_{C,AB} \tag{2.46}$$

Here rate equations for the change in concentration of each species have been constructed. For a simple system like the one above the equations can be solved analytically. However for more complicated systems with many different species and processes it is necessary to evaluate the differential equations computationally. The concentration of each species is evaluated as the time is increased by a discrete time step value. Solving these ordinary differential equations simultaneously will determine the continuous kinetic evolution of each of the different species A, B and C depending on the rates $k_{AB,C}$ and $k_{C,AB}$. This method is based on only the relative concentrations of the species. All spatial information, such as their diffusion and effects present in the limit of high concentrations are neglected. In particular, the rate constants are independent on the species environment; this approximation breaks down at large concentrations. Therefore solving rate equations will only yield an accurate representation of the kinetics in the limit of a low concentration of species, considering that their diffusion is fast enough not to limit any of the bimolecular reactions.

The rate constants can be calculated assuming harmonic transition state theory (Eq. (2.43)) and using the NEB method to calculate the relevant energy barriers e.g. $E_{AB,C}$. Based on hTST the pre-exponential factor ν in all cases will be approximately $10^{12} - 10^{13}$ s^{-1}.

2.7.3 Kinetic Monte Carlo

Kinetic Monte Carlo simulations differ from the rate equation approach described above in that instead of considering continuous concentrations of species that change in turn with time, the transition events are discretised. In this case instead of uniformly increasing the time, the time that has passed before a single transition can occur is calculated and then the system is evolved. For a system in state i there will be a rate constant that describes the probability of escape to any other state, k_{tot}. The probability, p_{ii} that the system remains in the same state i at time t is,

$$p_{ii}(t) = \exp(-k_{tot}t) \tag{2.47}$$

and hence the escape probability to have escaped after a time t' is $1 - p_{ii}(t')$. If the system was then to make an instantaneous transition to a state j the probability would be equal to the ratio of the rate constants for the transition to j and the total of all transitions, k_{ij}/k_{tot}. If the system remains in state i for a time t and then instantaneously makes a transition to state j then the probability is given by:

$$p_{ij} = \frac{k_{ij}}{k_{tot}} \exp(-k_{tot}t) \tag{2.48}$$

The time for the transition to occur Δt can be found from (2.47), assuming the probability can be considered as a random number (RN) between 0 and 1:

$$\Delta t = -\ln(RN)\frac{1}{k_{tot}} \tag{2.49}$$

The escape rate for a system in state i is formed of the sum of the rate of all possible transitions, $k_{tot} = \sum_j k_{ij}$.

The Bortz, Kalos and Lebowitz (BKL) algorithm is used to determine which transition is successful [29]. The rate constants for each transition are ordered into a "stack" and joined end to end. Each has a length equal to their value. The total length of the "stack" will therefore be equal to k_{tot}. Drawing a random number between 0 and k_{tot} and selecting the transition that this number lands on ensures that transitions are selected with the correct probability. Only one random number is required for this selecting process. The chosen process p' will then fulfil the criteria,

$$\sum_{p=1}^{p'} k_{ip} \geq RN \times k_{tot} \geq \sum_{p=1}^{p'-1} k_{ip} \tag{2.50}$$

where the rates for each process p out of a state i are summed. The BKL algorithm is illustrated in Fig. 2.6.

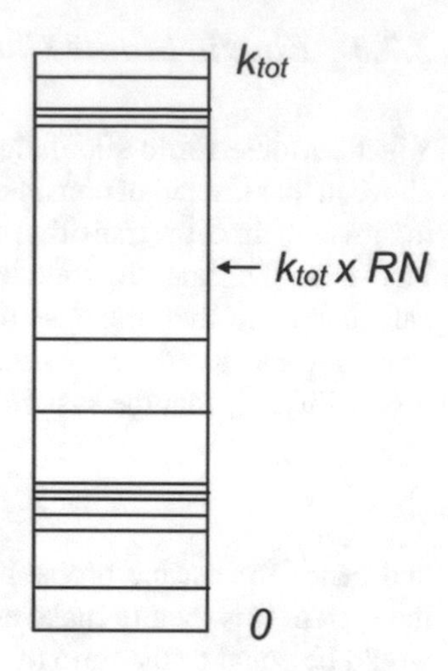

Fig. 2.6 The BKL algorithm for choosing a process in kMC. The rate constants are stacked into a list, where the height of each process corresponds to its rate constant, and the total height of the stack is k_{tot}. A random number between 0 and 1 is drawn and multiplied by k_{tot} in order to chose a process with probability proportional to its rate constant

Once the transition has been chosen and the time step calculated the system is updated with the new move. A new list of transitions is generated based on the new state and the process is then repeated.

2.7.3.1 Time-Dependent kMC

The kMC algorithm outlined above assumes that the rate constants are independent of the time. However for simulations of the thermal evolution of a system, the change in temperature will naturally affect the rates, which will mean that calculation of the time step will be time-dependent. Calculating Δt using Eq. (2.49) does not work, since k_{tot} at time t will be different from k_{tot} at $t \to t + \Delta t$. Instead the time step has to be calculated taking into account the time dependence of the rates [30]. For the usual system in a state i the probability of the system being still in state i after a time dt is:

$$p_i(t + dt) = p_i(t)[1 - k_{tot}(t)dt] \tag{2.51}$$

where $k_{tot}(t)$ is the time dependent rate. Rearranging this and solving the differential equation gives:

$$\frac{dp_i(t)}{dt} = -k_{tot}(t)p_i(t) \tag{2.52}$$

$$\ln p_i(t + \Delta t) = -\int_t^{t'+\Delta t} k_{tot}(\tau)d\tau \tag{2.53}$$

where Δt is the time which passes before the system can escape which is calculated from the time t of the previous step. For the time independent case, the right hand side simply becomes $k_{tot}\Delta t$ and Eq. (2.49) is recovered.

For time dependence, the time until escape, Δt, can be calculated from the integral,

$$-\ln(RN) = \int_t^{t+\Delta t} k_{tot}(\tau)d\tau \tag{2.54}$$

where RN is a random number between 0 and 1. Finding Δt then involves finding the upper limit of the integral over time. To do this the integral can be evaluated iteratively between t and $t + nt_i$, where $t_i << \Delta t$ and n is a positive integer so that the upper limit is increased gradually. Δt is found when the integral becomes greater than $-\ln(RN)$. This is when Δt has increased such that the probability of escape is certain.

2.7.4 Lattice-Based kMC

Another main difference between the rate equation approach and kMC is that spatial effects are included. In the case of reactions occurring on a surface simulations are carried out on a two-dimensional grid where the adsorbed molecules occupy grid sites. Molecules may diffuse between grid sites of the lattice during the simulation. Bimolecular reactions can only occur if the reacting molecules are located near to each other on the grid. This is dependent on whether the molecules can easily diffuse to meet each other. Likewise unimolecular decomposition reactions are only possible if there is the correct amount of empty space on the grid such that the product species can fit after a reaction. These reactions may become limited at high coverages where many sites are already occupied. For convenience, the simulations are run on a grid assuming that all species of interest can only occupy the grid vertices.

Local interactions between molecules can also be included with lattice-based kMC. The simplest way of achieving this is to consider nearest-neighbour interactions. For each different combination of neighbouring species the rate constant for reactions can be calculated depending on the location and the species of the nearest neighbour. For complicated systems with many different species this is unfeasible since there will be many different rates to calculate.

2.8 Molecular Dynamics

Molecular dynamics simulations are used to simply evolve the dynamical motion of a system of atoms or molecules over an incremental time interval. The motion of the a system of N atoms is simulated by evolving the Newtonian equations of motion at each time step dt. For a force F_i acting on atom i the motion will evolve with classical mechanics according to:

$$F_i = m_i \dot{v}_i \quad v_i = \dot{r}_i \tag{2.55}$$

The forces required for solving the equations of motion are derived from the atomic potential energy field. This may be determined using density functional theory or classically with empirical potentials. The equations of motion (2.55) are solved numerically for a finite time step dt by integration using, for example, the velocity Verlet algorithm [31]. The size of the time step must be of the order of femtoseconds in order to resolve atomic vibrations. This limits standard MD simulations to a total simulation time of microseconds when using empirical potentials and to picosecond when using DFT calculated forces.

For MD calculations different statistical ensembles can be used depending on the quantities of interest in the system. In the microcanonicial ensemble (or NVE) the number of particles N, the volume V, and the total energy E are constrained. In this ensemble the exchange between kinetic energy and potential energy of the atoms can be followed, however the total energy is constant.

In some scenarios the canonical ensemble (or NVT) is more suitable as the temperature T may be constrained, but the pressure is allowed to vary. This is useful for replicating experimental scenarios where high temperatures may be required. It is also practical for dissipation of energy through a system. The size of the system itself will affect the size of the temperature fluctuations.

2.8.1 Canonical Ensemble: NVT

To maintain the correct temperature in the canonical ensemble a thermostat needs to be applied to all or some of the atoms in the system. This type of simulation corresponds to the one in which a fragment of the whole system is attached to the rest of it (very large), which is treated as a heat bath at constant T, and hence adding or removing energy from the boundaries of the system when necessary can be facilitated. When doing this it is important to preserve the correct thermodynamics of the system and ensure that the dynamics is realistic.

2.8.1.1 Velocity Rescaling and the Berendsen Thermostat

The simplest way to ensure that the system temperature is constrained to the desired value is to rescale the velocity of the atoms. In the velocity rescaling method the velocity is scaled at each time step in order to adjust the temperature of the system.

$$\mathbf{v}_{new} = \lambda \mathbf{v}, \quad \lambda = \sqrt{\frac{T}{T_{inst}}}, \quad T_{inst} = \frac{2E_{kin}}{3k_B} \tag{2.56}$$

This however does not capture energy fluctuations in the system and the correct thermodynamics is lost.

The Berendsen thermostat also involves rescaling the average velocity of the atoms in order to adjust to the correct temperature, however instead of rescaling at every time step the rescaling takes place with a time scale for the heat transfer, τ, via [32]

$$\lambda^2 = 1 + \frac{\delta t}{\tau}\left(\frac{T}{T_{inst}} - 1\right). \tag{2.57}$$

As with velocity rescaling the Berendsen thermostat also fails to capture the correct thermodynamics.

2.8.1.2 Andersen Thermostat

In the Andersen thermostat the system is coupled to a heat bath, which allows effectively random stochastic forces to be applied to particles that are selected randomly [33]. The selected particle collides with the heat bath over a certain timescale and a new velocity is applied to it. The velocity is drawn from a Maxwell–Boltzmann distribution corresponding to the thermostat temperature. In the limit of an infinitely long simulation with many instances of velocity changes the canonical ensemble can be recovered. However once the velocity of the chosen particle is reassigned it will have lost all memory of its previous dynamics.

2.8.1.3 Nosé–Hoover Thermostat

In the Nosé–Hoover thermostat the temperature is regulated by the addition of an extra artificial degree of freedom, s to the equations of motion [34, 35]. This acts as a heat reservoir to exchange thermal energy with the system. The extra degree of freedom and the corresponding Hamiltonian are selected in such a way that the combined system is microcanonical; however, when the extra degree of freedom is integrated over, the system is exactly canonical. The Hamiltonian becomes:

$$H_{Nose} = \sum_{i=1}^{N} \frac{\mathbf{p_i}}{2m_i s^2} + U(\mathbf{r}^N) + \frac{p_s^2}{2Q} + nk_B T \ln s \tag{2.58}$$

where Q is an effective mass related to the additional degree of freedom s and n is the number of degrees of freedom, $n = 3N + 1$. It then follows that the momentum of s, p_s is equal to $Q\dot{s}$. The third and forth terms in Eq. (2.58) are related to the kinetic and "potential" energy of the artificial degree of freedom, respectively. The choice of the effective mass Q is important as this controls the coupling of the system to the reservoir. If Q is too large the coupling with the bath is reduced and the correct temperature will only be obtained after a long time. If Q is too small the temperature will fluctuate appreciably.

2.8.2 *Langevin Thermostat*

In order to ensure realistic dynamics of the system the Langevin thermostat can be used. Here the motion of slow particles throughout a continuum of fast particles is considered [36]. The small particles cause friction on the larger particles and also have some kinetic energy that transfers some energy to the larger particles. Correspondingly a friction term and force driving stochastic term $\mathbf{F}_i$ are added to the equations of motion for the large particles:

$$m_i \dot{\mathbf{v}}_i = -\frac{\partial U(\mathbf{r}^{\mathrm{N}})}{\partial \mathbf{r}_i} - \gamma m_i \mathbf{v}_i + \mathbf{F}_i \tag{2.59}$$

The friction term depends on the damping parameter γ. The driving force term must be balanced by the friction, so that the canonical ensemble is recovered. The force is chosen randomly within a Gaussian distribution with a dispersion σ_i given by [37]:

$$\sigma_i = \sqrt{2\gamma m_i k_B T / \Delta t} \tag{2.60}$$

where Δt is the MD time step. The exact dynamics can be determined providing the choice for γ is sensible. A method for determining a value for the damping parameter is detailed below.

2.8.2.1 Determining the Damping Parameter

The value of the damping parameter in the Langevin thermostat has to be chosen so that the correct dynamics of the system is obtained. To do this it is possible to determine the phonon DOS of the system from the molecular dynamics with the thermostat applied to some of the atoms and then compare this to the phonon DOS as determined from lattice dynamics. The phonon DOS can be found from an MD simulation by calculating the velocity autocorrelation function of a selection of NVE atoms located within a system where the thermostat is applied to a different selection of NVT atoms. In this setup kinetic energy will be added and removed to the NVE atoms from the NVT atoms, while the NVE atoms maintain the proper dynamics. Taking the Fourier transform of the velocity autocorrelation function returns the phonon DOS.

The velocity autocorrelation function (VACF) determines how correlated the velocity of a system is at two separate times. For a velocity of a particle at a time t and at future time $t + \tau$ the correlation function will decay as τ is increased and the two velocities become further apart in time. This is illustrated in Fig. 2.7.

The VACF, $G(\tau)$ is given by

$$G(\tau) = \sum_i \langle v_i(t_0) v_i(t_0 + \tau) \rangle_{t_0} = \frac{1}{N} \sum_{i=1}^{N} \frac{1}{t_{max}} \sum_{t_0=1}^{t_{max}} v_i(t_0) v_i(t_0 + \tau), \tag{2.61}$$

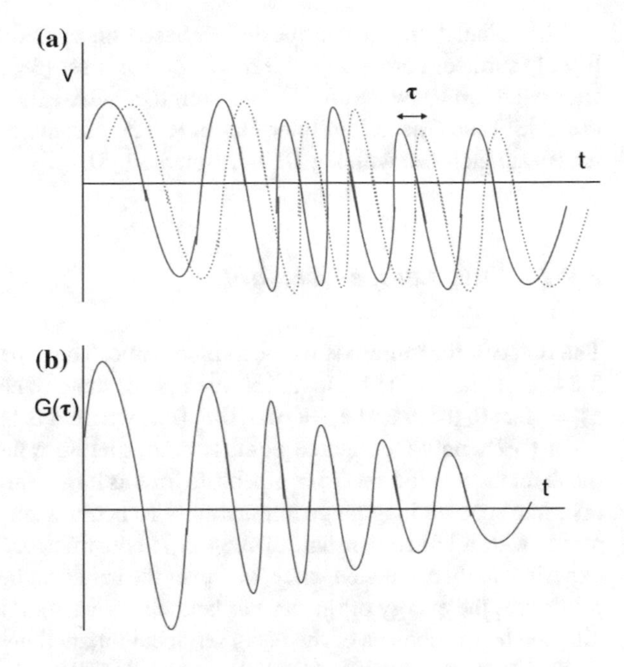

Fig. 2.7 **a** The velocity of a particle at t_0, $v_i(t_0)$ and shifted by τ, $v_i(t_0 + \tau)$ and **b** the corresponding velocity autocorrelation function $G(\tau)$

where v_i is the velocity of particle i. In Eq. (2.61) the velocity correlation function of each particle is found as τ is increased linearly until the end of the simulation time is reached t_{max}. A particle average is then taken by summing the individual velocity correlation functions and then dividing by the number of particles N.

By taking the Fourier transform of $G(\tau)$ the phonon DOS is recovered. By varying γ and monitoring the changes in the phonon DOS an optimum value for γ can be chosen where the calculated phonon DOS matches the features of a reference phonon DOS, as determined from lattice dynamics simulations.

2.9 Ir(111) Surface Parameterisation

The first step in any DFT calculation is to parameterise the system. For a bulk crystalline system this means determining the correct lattice constant given the basis set, basis set cut off energy and the pseudopotentials. Since periodic boundary conditions are applied, to construct a surface slab a vacuum gap must be introduced to separate adjacent layers. Parameterising the surface set up requires determining a suitable value for the vacuum gap, as well as the number of layers required to approximate the surface. More layers are favourable to get a more accurate result; however this increases the number of atoms in the simulation, and slows down calculations.

The calculations in this thesis are based on epitaxial graphene growth on the Ir(111) surface. For iridium the m-DVZP basis set [38], with a plane wave cut off energy of 300 Ry were used, along with the GGA-PBE exchange-correlation functional [39] and Goedecker-Teter-Hutter (GTH) pseudopotentials [19]. Van der Waals forces are included with the DFT-D3 method [8].

2.9.1 Bulk Lattice Constant

The unit cell for iridium is face-centered cubic. The experimental lattice constant is 3.84 Å and hence the basic lattice vectors for the unit cell are: $\mathbf{a}_1 = (0, 0.5, 0.5)a$, $\mathbf{a}_2 = (0.5, 0, 0.5)a$ and $\mathbf{a}_3 = (0.5, 0.5, 0)a$, where a is the lattice constant.

In CP2K only the gamma point is used, therefore there is not a well defined k mesh for the integration. To compensate for this large real-space cells must be used to take into account long range interactions. To begin with, a $4 \times 4 \times 4$ cell of iridium atoms with a lattice constant of 3.84 Å is constructed. This lattice constant is the experimentally calculated value, however the actual value of the lattice constant that minimises the energy of the atomic lattice may vary slightly from this. This will be affected by the choices of the basis set, local interactions and pseudopotentials.

To determine the value of the lattice constant of the bulk Ir the unit cell is scaled and the energy calculated. By fitting a parabola to the results and finding the minimum, the optimum lattice constant value is found. Based on the results in Fig. 2.8 the lattice constant for iridium is 3.81 Å. This is very close to the experimental value, and suggests that the calculation set up accurately describes the system. Correspondingly, the Ir lattice is rescaled to this calculated lattice constant value.

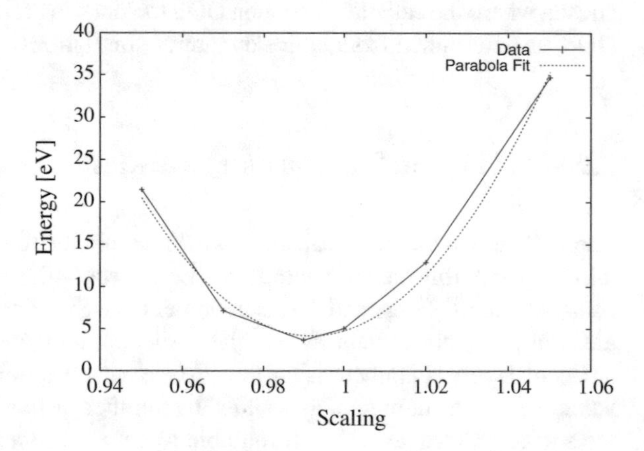

Fig. 2.8 The total energy of the Ir bulk lattice for different scaling applied to the primitive unit cell. The *dotted line* is an interpolation of the calculated energies with a parabola

2.9.2 Ir(111) Surface

From the Ir bulk an Ir(111) surface slab can be created by dissecting along the (111) Miller indices of the unit cell and introducing a vacuum gap to produce slabs that are separated from each other in each unit cell. Since CP2K requires a large unit cell size, and to ensure molecules placed on the surface are well separated an 8×8 unit cell is used for a single Ir(111) layer.

Creating a surface with a large number of layers gives the most accurate representation, however this is too computationally expensive. Therefore only a few layers may be used, however they must still be able to represent the properties and behaviour of a surface with many layers. These include the surface energy, the electronic properties (DOS and band structure) and its interaction with any surface species placed onto it.

First the change in surface energy with the number of layers is determined. The surface energy per atom is defined as:

$$E_{surf} = \frac{1}{2N}(E_{slab} - NE_{bulk}), \tag{2.62}$$

where E_{slab} is the total energy of the slab, E_{bulk} is the energy of a single bulk atom (in the case of Ir) and N is the total number of atoms in the slab. In this calculation both the upper and the lower surfaces are allowed to relax.

As the number of layers increases this energy will converge when the top and bottom surfaces are far enough apart and do not interact with each other. This gives an idea of how many layers should be used. The surface energy for the Ir(111) surface with varying number of layers is shown in Fig. 2.9. The vacuum gap is greater than 15 Å.

For more than three layers the surface energy varies by less than 0.05 eV/Atom. Therefore more than two layers should be used in order to get an accurate result (less

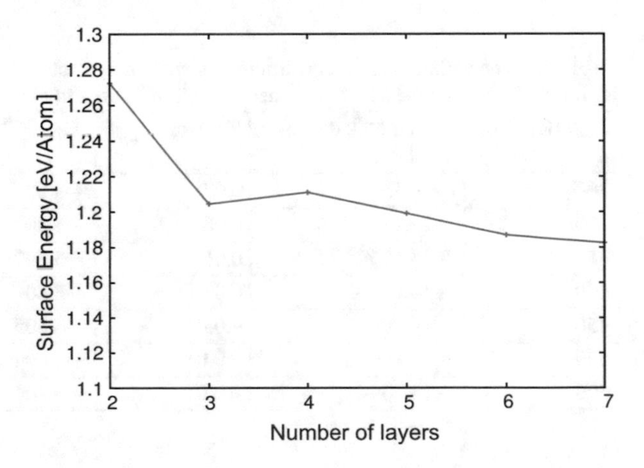

Fig. 2.9 The surface energy per surface atom of an 8×8 Ir(111) surface cell for different numbers of layers

than 0.05 eV error). Even with only two layers the energy is still within 0.1 eV of its converged value. As adding an extra layer to the system requires 64 more atoms it is computationally expensive to consider many layers. For the surface calculations in this thesis four layers are used unless stated. Two layers were used initially for expensive calculations such as NEBs. To obtain more accurate values for the energy barriers this was increased to up to 7 layers, where necessary. This is detailed in the Appendix A. In all cases the vacuum gap is greater than 15 Å.

In order to preserve the bulk geometry in the surface slab a number of the bottom layers must be fixed to the bulk lattice constant, while the upper layers are free to relax. The number of layers which are fixed to the Ir bulk geometry may be varied. The system must have enough of these to be similar to the bulk, but near the top of the surface some layers must be allowed to relax so that it can behave correctly near the vacuum gap so that it can interact with any molecules which may be added to it.

2.9.3 Plane Wave Cutoff Energy

The plane wave cutoff energy E_{cut} dictates the cut off for the largest plane waves vectors G_{max} where $E_{cut} = \hbar^2 G_{max}^2/2m$, where m is the electron mass. Choosing a large enough value is crucial in order to include enough plane waves to give accurate results. The convergence of the total energy of the system can be monitored as the cutoff energy is increased until the energy converges.

Energy differences will converge with E_{cut} much faster than total energies. Since only the relative energy differences are of interest (e.g. adsorption energies, energy barriers, binding energies and formation energies) it is possible to take a lower value for the energy cutoff. The calculated formation energy for a vacancy defect in the iridium bulk and the adsorption energy for a CH_2CH molecule on the Ir(111) surface are shown in Table 2.1 for a variety of energy cutoffs. To determine the formation energy of the vacancy defect in the bulk Ir the energy difference between the energy

Table 2.1 The variation of energy differences with energy cutoff E_{cut} for the formation energy of an Ir bulk defect and the adsorption energy of CH_2CH on a Ir(111) surface with 3 layers

E_{cut} [Ry]	E_f (Ir bulk defect) [eV]	% Error	E_{ads} (CH_2CH) [eV]	% Error
200	−14.154	0.07	−3.802	0.05
300	−14.155	0.06	−3.802	0.05
350	−14.155	0.06	−3.802	0.05
400	−14.161	0.02	−3.803	0.02
450	−14.164	<0.01	−3.805	0.02
500	−14.164	<0.01	−3.804	<0.01
600	−14.164	<0.01	−3.804	<0.01

of the defective cell E_{defect} plus the energy of a single Ir atom E_{atom} and the energy of defect-free cell E_{bulk} is calculated:

$$E_f^{defect} = [E_{defect} + E_{atom}] - E_{bulk} \tag{2.63}$$

The adsorption energy of the CH_2CH molecule on the Ir(111) surface is given by:

$$E_{ads} = E_{Ir+CH_2CH} - [E_{surf} + E_{CH_2CH}] \tag{2.64}$$

In each case the geometries are relaxed. Overall the error in the energies due to the cutoff is very small (less than 1%). Almost any of the tried values could be taken for the energy cutoff. In the calculations of this thesis a value of 300 Ry is used.

References

1. M. Born, R. Oppenheimer, Zur quantentheorie der molekeln. Annalen der Physik (Leipzig), **84**, 457 (1927)
2. P. Hohenberg, W. Kohn, Inhomogeneous electron gas. Phys. Rev. **136**(3B), B864–B871 (1964)
3. W. Kohn, L.J. Sham, Self-consistent equations including exchange and correlation effects. Phys. Rev. **140**(4A), A1133–A1138 (1965)
4. D.M. Ceperley, B.J. Alder, Ground state of the electronic gas by a stochastic method. Phys. Rev. Lett. **45**(7), 566–569 (1980)
5. J.P. Perdew, J.A. Chevary, S.H. Vosko, K.A. Jackson, M.R. Pederson, D.J. Singh, C. Fiolhais, Atoms, molecules, solids, and surfaces - applications of the generalized gradient approximation for exchange and correlation. Phys. Rev. B **46**(11), 6671–6687 (1992)
6. A.D. Becke, Density-functional exchange-energy approximation with correct asymptotic behavior. Phys. Rev. A **38**(6), 3098–3100 (1988)
7. S. Grimme, Semiempirical GGA-type density functional constructed with a long-range dispersion correction. J. Comput. Chem. **27**(15), 1787–1799 (2006)
8. S. Grimme, J. Antony, S. Ehrlich, H. Krieg, A consistent and accurate ab initio parametrization of density functional dispersion correction (DFT-D) for the 94 elements H-Pu. J. Chem. Phys. **132**(15), 154104 (2010)
9. A. Tkatchenko, M. Scheffler, Accurate molecular van der waals interactions from ground-state electron density and free-atom reference data. Phys. Rev. Lett. **102**, 073005 (2009)
10. D.C. Langreth, M. Dion, H. Rydberg, E. Schröder, P. Hyldgaard, B.I. Lundqvist, Van der waals density functional theory with applications. Int. J. Quantum Chem. **101**(5), 599–610 (2005)
11. M. Dion, H. Rydberg, E. Schröder, D.C. Langreth, B.I. Lundqvist, Van der Waals density functional for general geometries. Phys. Rev. Lett. **92**, 246401 (2004)
12. J. Bamidele, J. Brndiar, A. Gulans, L. Kantorovich, I. Štich, Critical importance of van der waals stabilization in strongly chemically bonded surfaces: Cu(110):O. J. Chem. Theory Comput. **9**(12), 5578–5584 (2013). PMID: 26592291
13. L. Kantorovich, *Quantum Theory of the Solid State: An Introduction* (Kluwer, 2004)
14. S.F. Boys, F. Bernardi, The calculation of small molecular interactions by the differences of separate total energies. Some procedures with reduced errors. Molecul. Phys. **19**, 553–566 (1970)
15. G. Kresse, J. Furthmüller, Efficiency of ab-initio total energy calculations for metals and semiconductors using a plane-wave basis set. Comput. Mater. Sci. **6**(6), 15–50 (1996)
16. G. Kresse, J. Furthmüller, Efficient iterative schemes for ab initio total-energy calculations using a plane-wave basis set. Phys. Rev. B **54**(16), 11169–11186 (1996)

17. J. Hutter, M. Iannuzzi, F. Schiffmann, J. VandeVondele, cp2k: atomistic simulations of condensed matter systems. Wiley Interdiscip. Rev. Comput. Molecul. Sci. **4**(1), 15–25 (2014)
18. H.J. Monkhorst, J.D. Pack, Special points for brillouin-zone integrations. Phys. Rev. B **13**, 5188–5192 (1976)
19. S. Goedecker, M. Teter, J. Hutter, Separable dual-space gaussian pseudopotentials. Phys. Rev. B **54**, 1703–1710 (1996)
20. G. Henkelman, B.P. Uberuaga, H. Jónsson, A climbing image nudged elastic band method for finding saddle points and minimum energy paths. J. Chem. Phys. **113**, 9901 (2000)
21. G. Henkelman, H. Jónsson, Improved tangent estimate in the nudged elastic band method for finding minimum energy paths and saddle points. J. Chem. Phys. **113**, 9978 (2000)
22. H. Jónsson, G. Mills, K.W. Jacobsen, Nudged elastic band method for finding minimum energy paths of transitions, in *The Book "Classical and Quantum Dynamics in Condensed Phase Simulations"* (1999), p. 385
23. G.H. Vineyard, Frequency factors and isotope effects in solid state rate processes. J. Phys. Chem. Solids **3**(1), 121–127 (1957)
24. K. Kunc, R.M. Martin, Ab Initio force constants of gaas: a new approach to calculation of phonons and dielectric properties. Phys. Rev. Lett. **48**, 406–409 (1982)
25. L. Köhler, G. Kresse, Density functional study of CO on Rh(111). Phys. Rev. B **70**, 165405 (2004)
26. A.F. Voter, *Radiation effects in solids*. NATO Publishing unitHandbook of Material Modeling, Part A. Methods (Springer, NATO Publishing Unit, Dordrecht 2005)
27. A.F. Voter, J.D. Doll, Transition state theory description of surface selfdiffusion: comparison with classical trajectory results. J. Chem. Phys. **80**(11) (1984)
28. P. Hänggi, P. Talkner, M. Borkovec, Reaction-rate theory: fifty years after kramers. Rev. Modern Phys. **62**, 251–341 (1990)
29. A.B. Bortz, M.H. Kalos, J.L. Lebowitz, A new algorithm for monte carlo simulation of ising spin systems. J. Comput. Phys. **17**(1), 10–18 (1975)
30. A. Prados, J.J. Brey, B. Sánchez-Rey, A dynamical monte carlo algorithm for master equations with time-dependent transition rates. J. Stat. Phys. **89**(3), 709–734 (1997)
31. D. Frenkel, B. Smit, *Understanding Molecular Simulation, From Algorithms to Applications* (Aced. Press, 2002)
32. H.J.C. Berendsen, J.P.M. Postma, W.F. van Gunsteren, A. DiNola, J.R. Haak. Molecular dynamics with coupling to an external bath. J. Chem. Phys. **81**(8) (1984)
33. H.C. Andersen, Molecular dynamics simulations at constant pressure and/or temperature. J. Chem. Phys. **72**(4) (1980)
34. S. Nosé, A unified formulation of the constant temperature molecular dynamics methods. J. Chem. Phys. **81**(1) (1984)
35. W.G. Hoover, Canonical dynamics: equilibrium phase-space distributions. Phys. Rev. A **31**, 1695–1697 (1985)
36. T. Schneider, E. Stoll, Molecular-dynamics study of a three-dimensional one-component model for distortive phase transitions. Phys. Rev. B **17**, 1302–1322 (1978)
37. B. Dünweg, P. Wolfgang, Brownian dynamics simulations without gaussian random numbers. Int. J. Modern Phys. C **02**(03), 817–827 (1991)
38. J. VandeVondele, J. Hutter, Gaussian basis sets for accurate calculations on molecular systems in gas and condensed phases. J. Chem. Phys. **127**(11), 114105 (2007)
39. J.P. Perdew, K. Burke, M. Ernzerhof, Generalized gradient approximation made simple. Phys. Rev. Lett. **77**, 3865 (1996)

Chapter 3
Producing a Source of Carbon: Hydrocarbon Decomposition

In this chapter the decomposition mechanism for ethylene on the Ir(111) surface will be determined. This is motivated by the early stages of epitaxial graphene growth where hydrocarbons are deposited onto a substrate, and then by heating the system they decompose to form carbon based species that go on to form graphene. This study is based on temperature programmed growth, where the deposition occurs at room temperature, prior to a heating phase where the temperature may be ramped up with a constant heating rate, or increased instantly to a constant value.

The theoretical results of this section are supported with the experimental work by the group of Prof. A. Baraldi at the University of Trieste and the Eletta Syncotron Trieste. Temperature programmed-XPS experiments were performed in order to image the C 1s core level spectra under growth conditions. To assist in the identification of the spectra binding energy calculations of the C atoms in each molecule are performed.

3.1 Theoretical Method Outline

The method is based on previous studies [1], as mentioned in Sect. 1.2.1. To understand the all possible decomposition processes starting from the ethylene molecule first a complete reaction scheme is devised. Based on this reaction scheme the energy barriers associated with each reaction are calculated using the nudged elastic band method. By finding the initial and final states of each reaction, the transition state, corresponding to the saddle point, can be determined and used to calculate the energy barrier. For these calculations first the lowest energy configurations for each hydrocarbon species need to be found.

Once the energy barriers for each reaction are calculated they can be used to determine the reaction rates using the Arrhenius equation,

$$R_{i,j} = \nu_{i,j} e^{-E_{i,j}/k_B T} \tag{3.1}$$

H.A. Tetlow, *Theoretical Modeling of Epitaxial Graphene Growth on the Ir(111) Surface*, Springer Theses, DOI 10.1007/978-3-319-65972-5_3

where $E_{i,j}$ are the forward (i) and reverse (j) barriers for a reaction respectively, k_B is the Boltzmann constant, and T is the temperature. The pre-exponential factors, $\nu_{i,j}$ are typically of the order of 10^{13}s^{-1}, but a more precise value can be determined from vibrational frequency calculations (as given in Eq. 2.43). Finally the kinetics of the decomposition process is considered, and the thermal evolution of species on the surface is calculated. This is done first by solving rate equations, and then with a kinetic Monte Carlo simulation, where spatial effects can be included.

3.2 Decomposition Reaction Scheme

The reaction scheme in Fig. 3.1 is formed by considering all possible reactions starting from ethylene CH_2CH_2, and connecting all species $C_{(1,2)}H_{(1-6)}$ containing two C atoms or less. The various reaction processes consist of hydrogenation (adding

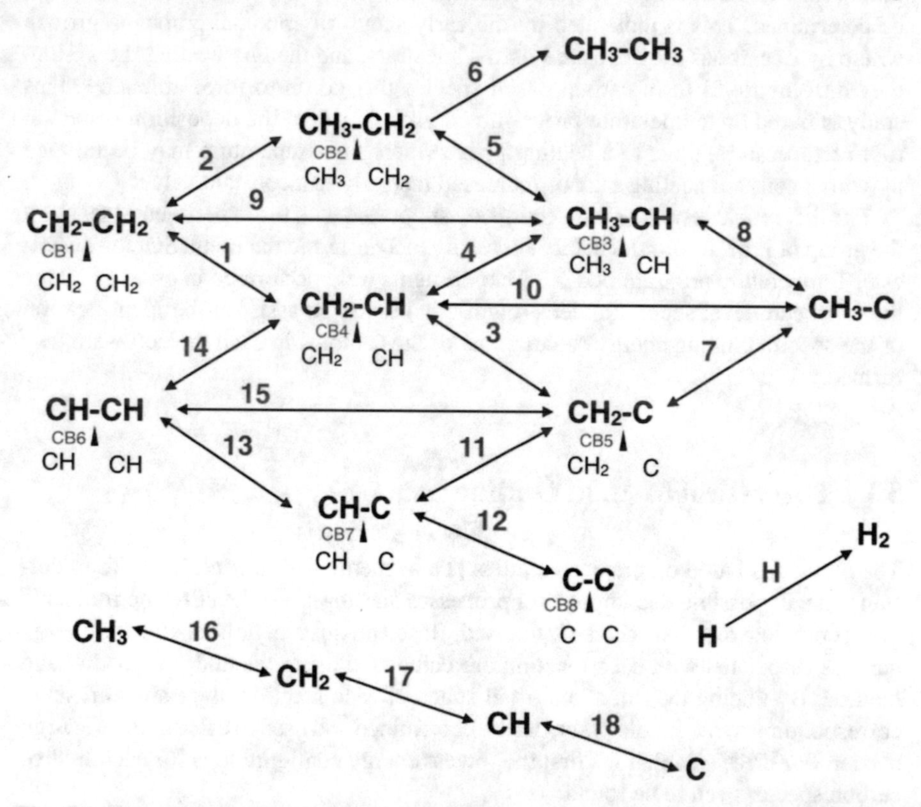

Fig. 3.1 The reaction scheme connecting all hydrocarbon species $C_{(1,2)}H_{(1-6)}$. The reaction processes consist of hydrogenation reactions (*upward diagonal arrows*), dehydrogenation reactions (*downward diagonal arrows*) and isomerisation reactions (*horizontal arrows*). These are numbered from 1 to 18. C-C breaking reactions, and the reverse process, C-C recombination are indicated by the small vertical arrows and are numbered from CB1 to CB8. Two hydrogen atoms may also form H_2 and desorb (symbolised by H). Reproduced from Ref. [2] with permission from the PCCP Owner Societies

surface adsorbed H atoms), dehydrogenation reactions (removing H atom to the surface) and isomerisation reactions (repositioning of the H atom from one C atom to the other), as well as breaking of the C-C bond in C_2H_m species, and C-C bond reformation from two CH_m fragments. Hydrogen monomers may also be lost from the surface by forming molecular H_2. This process must be included since it affects the concentration of H atoms on the surface, which will lessen the likelihood of hydrogenation reactions.

3.3 Hydrocarbon Species

In order to determine the energetics of each reaction the corresponding minima for the initial and final reaction states need to be found. This requires determining the optimised geometries of all the adsorbed hydrocarbons. Each hydrocarbon species is placed onto the Ir(111) surface and then its geometry is relaxed using DFT. In many cases multiple stable geometries are found for each species. For the subsequent NEB calculations (and reaction rates) only the lowest energy configurations are used. The lowest energy geometries obtained for each species are shown in Fig. 3.2 along with their adsorption energies.

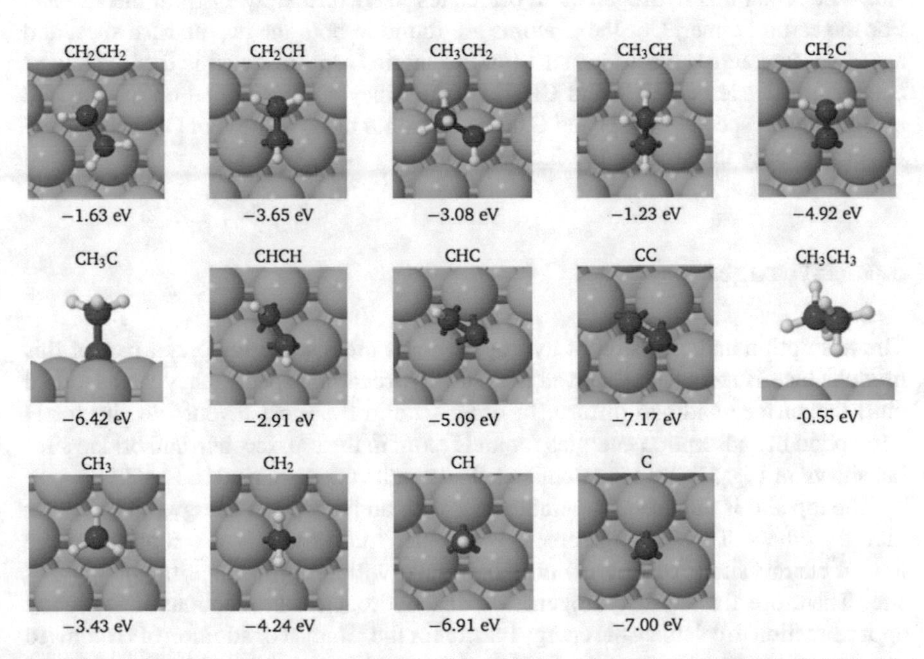

Fig. 3.2 The lowest energy geometries of the hydrocarbon species adsorbed onto the Ir(111) surface. The adsorption energies of each species is also included. Ir atoms are shown in blue, while C and H are *dark* and *light grey*, respectively. Reproduced from Ref. [2] with permission from the PCCP Owner Societies

The adsorption energies are calculated using

$$E_{ads} = E_{Ir+C_nH_m} - E_{Ir} - E_{C_nH_m} \tag{3.2}$$

where $E_{Ir+C_nH_m}$ is the energy of the molecule adsorbed on the surface, E_{Ir} is the energy of the bare Ir(111) surface and $E_{C_nH_m}$ is the energy of the molecule in the gas phase. For these calculations an 8×8 cell is used (to ensure the molecules do not interact across unit cells), with four Ir layers, and a vacuum gap of greater than 15 Å. The bottom two of the four layers are fixed to the bulk geometry. Specific details relating to the parameterisation of the Ir(111) surface for the DFT calculations can be found in Sect. 2.9.

It is found that all species, apart from CH_3CH_3 are highly stable on the Ir(111) surface. For ethylene (CH_2CH_2) the lowest energy configuration is obtained when the molecule is in the bridge position, between two Ir atoms. The species CH_2CH, CH_3CH, CH_3C and CH_2C are centred on either the fcc or hcp three-fold hollow sites. The difference in adsorption energy between attachment at fcc and hcp sites is negligible. For CH_3CH_2 the CH_2 group is attached to the top site, while the CH_3 is suspended away from the surface. CH_3CH_3 is weakly bound to the surface. The structure of CHCH has each of the CH groups in a bridge site. Compared to CHCH, in CHC the C atom without hydrogen shifts towards the other to rest in the hollow site. The remaining hydrogen atom orientates itself further away from the surface. For the carbon dimer, CC, the C atoms are found in both the hcp and fcc sites, and are connected across the bridge site. For the single C species we find that the lowest energy geometries for CH_3 and CH_2 are when they are positioned on the top and bridge sites, respectively. CH and C both relax onto the hcp site. For C this is 0.2 eV lower in energy than for the fcc site geometry.

3.4 Hydrogen

The adsorption and diffusion of hydrogen atoms are also an important part of this model. Once H atoms are removed from the hydrocarbon species they are adsorbed onto the surface and can diffuse. In order to find the most favourable site for H adsorption the adsorption energies for an H atom in the top, fcc, hcp and bridge sites (as shown in Fig. 3.3a) were calculated. The results are displayed in Fig. 3.3b.

The top site is the most favourable site, with an adsorption energy 0.3 eV lower than the others. This is in agreement with other studies which also report it as the lowest energy site [3]. H atoms on the surface will likely become trapped at this site. Therefore for the dehydrogenation and hydrogenation reactions it is a good approximation to calculate the energy barrier for the removal or addition of H atoms to or from the top site. For the final (initial) state in a dehydrogenation (hydrogenation) NEB energy barrier calculations the H atom will be relaxed on top site near the hydrocarbon species.

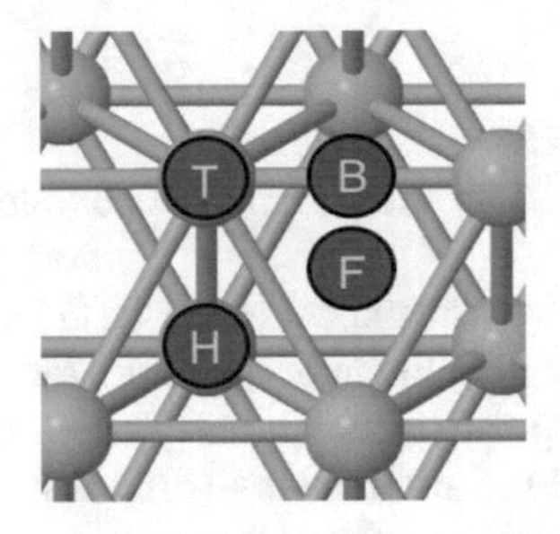

(a) The letters F, H, B and T show the positions of the fcc, hcp, bridge and top sites respectively.

Site	E_{ads} [eV]
bridge	-2.729
fcc	-2.750
hcp	-2.704
top	-3.054

(b) The absorption energy for H in different sites on the Ir(111) surface.

Fig. 3.3 The adsorption of H on the Ir(111) surface

Table 3.1 The forward and reverse barriers for hydrogen diffusion

Diffusion process	E_{for} [eV]	E_{back} [eV]
Bridge → fcc	≃0	0.015
Bridge → hcp	0.03	0.005
Bridge → top	0.02	0.27
Top → fcc	0.36	0.05

The mobility of H atoms on the surface will affect dehydrogenation and hydrogenation reactions. If the barrier for H diffusion is large with respect to the barrier for hydrogenation reactions then once hydrogen has been removed from a molecule to the surface it is more likely to recombine with the molecule rather than diffuse away from it. A large H diffusion barrier would also indicate that hydrogenation reactions could become limited by the mobility of the H atoms. Therefore the H diffusion barriers are calculated.

For the sites in Fig. 3.3 the H diffusion barrier was calculated from one site to another, along a linear path. The barriers are shown in Table 3.1. From examination of the four adsorption sites in Fig. 3.3 it can be seen that diffusion between some sites is equivalent to others. For example diffusion between an fcc and an hcp site is equivalent to fcc → bridge then bridge → hcp, or fcc → top then top → hcp depending on which pathway has the lowest energy barrier.

For diffusion from the top site to the hcp site, the pathway is unstable, such that H will prefer to diffuse via the bridge site. For diffusion between two top sites, i.e. for H atoms moving between its lowest energy adsorption site, the H will diffuse via a bridge site, with an energy barrier of 0.27 eV. The largest barrier for H diffusion from the top to the fcc site is 0.36 eV. This is small, especially when considering the high temperatures used in graphene growth. Therefore it should be expected that H atoms are highly mobile on the surface.

3.5 Photoemission Experiments

The photoemission experiments were performed by the group of A. Baraldi at the University of Trieste, Italy.

Photoemission experiments are performed by first exposing the Ir(111) surface to ethylene at 90 K. Then the system is heated with a linear temperature ramp of 1.5 Ks^{-1}. Over the course of the ramping the XPS spectra are acquired. These are shown in Fig. 3.4b–d for coverages of 0.05, 0.12 and 0.6 monolayers (ML) respectively. Here the coverage refers to the coverage of C atoms on the surface.

For all coverages at around 250, 380 and 500 K transitions in the peaks spectra are noticed corresponding to the evolution of the hydrocarbon species on the surface. Depending on the initial coverage the temperature of these transitions may vary, however overall the data are qualitatively the same, and hence the initial ethylene coverage does not greatly affect the evolution of the species on the surface. Above 800 K the spectrum narrows and moves towards a value of around 284 eV. This binding energy value is commonly associated with the production of graphene islands.

Figure 3.4a shows the coverage dependent results after the deposition at 90 K. Two peaks at 283.3 eV (a_1) and 283.8 eV (a_2) are observed upon the exposure to ethylene. As the exposure is increased another peak appears at 282.7 eV (b_1), accompanied by a peak at about 283 eV (b_2), which is hidden by the a_1 and a_2 peaks. When comparing this data with the temperature ramping spectra for the coverage of 0.6 ML (d) at 90 K these peaks are all present initially. However as the temperature is increased peaks b_1 and b_2 transform into a_1 and a_2. This suggests that at high coverages ethylene (peaks b_1, b_2) arrives at the surface, and partially dissociates into another species

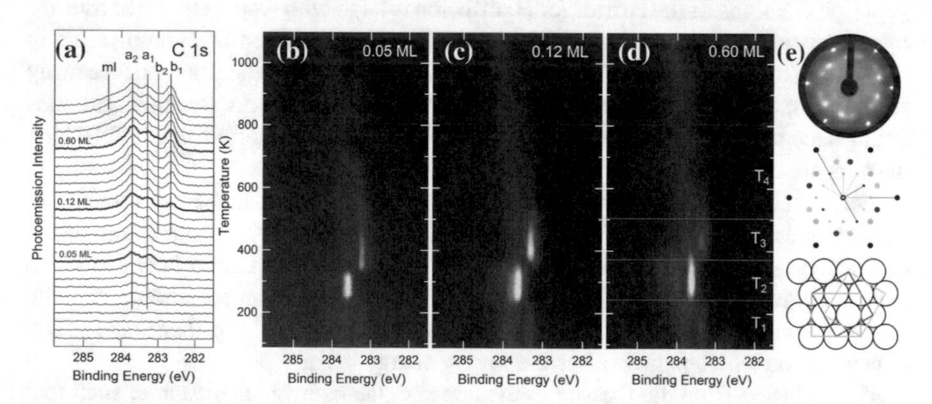

Fig. 3.4 **a** XPS C 1s spectra with increasing ethylene dosage at 90 K. Three different coverages are indicated by the red lines, 0.05 ML, 0.1 ML and 0.6 ML. **b–d** The temperature programmed XPS C 1s spectral series produced during the annealing from 90 K to high temperature after ethylene deposition for the three different ethylene coverages. Each horizontal cut in the figures corresponds to a single photoemission spectrum acquired at a given temperature and its intensity is plotted using a colour scale. Reproduced from Ref. [2] with permission from the PCCP Owner Societies

(peaks a_1, a_2). This process only completes once the temperature is increased to 200 K. This is unlike low coverage case where ethylene dissociates completely soon after deposition. This effect suggests that at high coverages saturation of the surface prevents ethylene fully dissociating initially.

3.5.1 Binding Energy Calculations

In order to help determine which hydrocarbon species correspond to the XPS spectra peaks, the C 1s core level binding energies for each species are calculated using DFT. The method for this is detailed in Sect. 2.6. The relative binding energies are presented in Fig. 3.5 along with pictures to illustrate which C atom the value relates

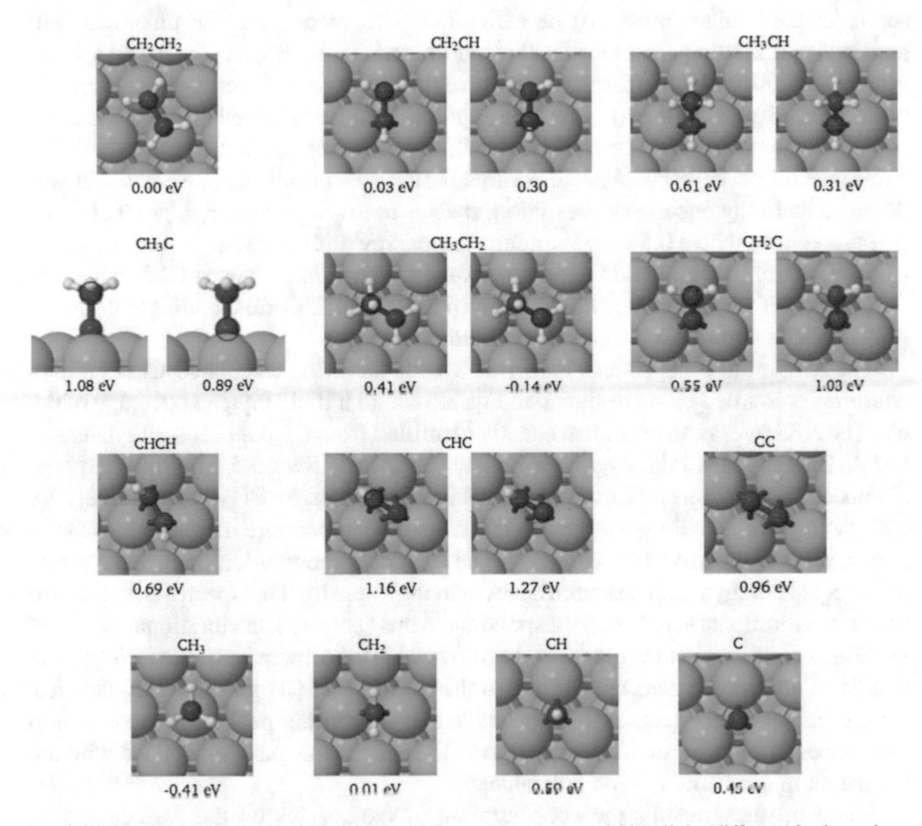

Fig. 3.5 The C 1s core level shift calculated for carbon atoms within all the different hydrocarbon species with respect to ethylene (before the shift to fit the XPS spectra). When there are two non-equivalent C atoms, two energies are given, corresponding to the *circled* atoms in the *left* and *right* panels. For species where the C atoms appear in the same environment, E_{CLS} is the same and, hence, only one value is given. Reproduced from Ref. [2] with permission from the PCCP Owner Societies

to. The values are listed with respect to the binding energy of the C atoms in ethylene adsorbed on the surface. For equivalent C atoms, (for example in CHCH) only one value for the binding energy is shown since the values for each atom should be equal.

The method used to calculate the core level shift does not give the exact values for the binding energies, but rather the relative energy difference between the different C environments. Hence in order to use this data to identify the XPS peaks, first all binding energies must be shifted by some value in order to align with the experimental results.

3.5.2 *Interpretation of XPS Data*

In order to use the core level binding energy calculations to interpret the XPS data, the calculated values must first be calibrated. This means that the unknown shift applied to all binding energies must be determined. To do this one (or more) species must be initially identified so that the peak(s) in the corresponding spectrum can be associated with it (them). Then the appropriate linear shift can be applied to all the DFT calculated core level shift (CLS) data so that the values align with the experimental peaks. Based on the photoemission spectra results in Sect. 3.5, it was deduced that ethylene dissociates into a species at low temperatures, producing the peaks a_1 and a_2 with a 0.5 eV separation. The energy difference between these peaks (0.5 eV) matches with the difference between the calculated values CLS values for CH_2C. Therefore CH_2C may be assigned to these peaks. Correspondingly a shift of +282.8 eV is applied to the calculated binding energies.

The temperature dependent XPS peaks with the fitted calculated CLS binding energies (coloured ticks) are shown in Fig. 3.6 for an initial ethylene coverage of 0.6 eV. The XPS peaks can be almost exactly identified from the data. Both ethylene and CH_2C are produced initially, as previously deduced in Sect. 3.5 from the coverage dependent data. Therefore ethylene quickly dissociates to CH_2C, most likely by completing two dehydrogenation reactions. As the temperature increases to 323 K a peak corresponding to CHCH appears. The XPS data shows additional peaks which are an equal energy width apart and have decaying intensity. This spectra is deduced to be due to vibrational satellites corresponding to this species. The vibrational modes of these are calculated in the Appendix A. Above 443 K the intensity of the CHCH peak decreases and more peaks appear due to the presence of CH and CH_3C molecules. As the temperature increases these peaks flatten and further peaks corresponding to the species C and C_2 can be identified. At 1133 K a narrow peak associated with the formation of graphene islands is produced.

Based on these results the concentration of the species on the surface can be deduced over the temperature range of the experiment. Details of the method for extracting the concentrations of each species from the XPS data are given in [2]. The experimentally deduced thermal evolution of species on the surface is shown in Fig. 3.7, based on the calculated relative BEs (and the corresponding energy shift) used to identify the species peaks.

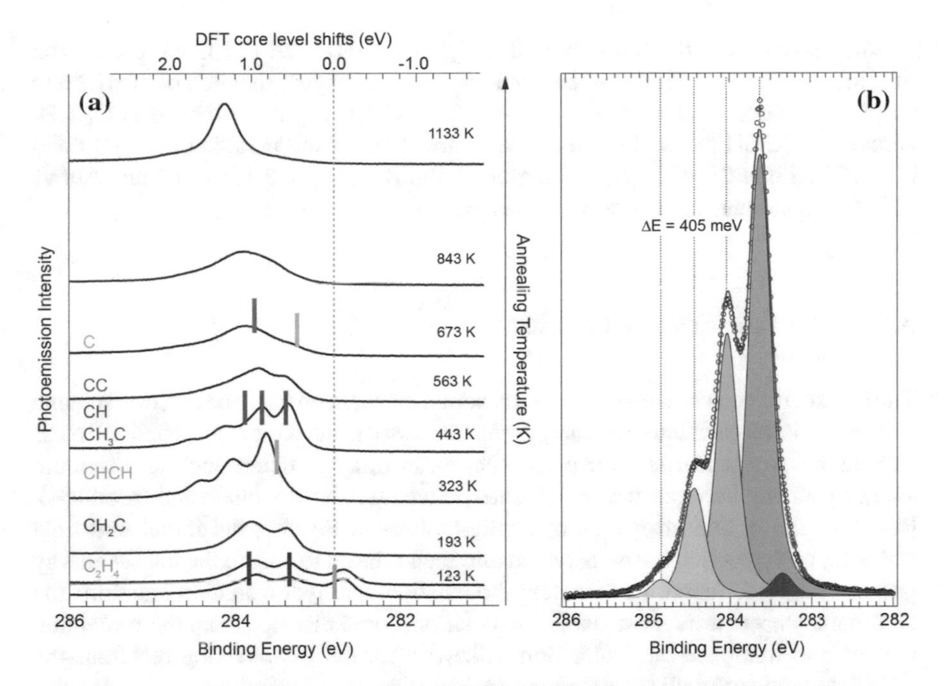

Fig. 3.6 **a** High-resolution XPS spectra acquired after deposition of ethylene at 90 K and subsequent annealing to the temperatures shown. The DFT calculated BEs (with the shift applied) for the species have been marked with colour ticks. **b** The spectrum of the system after annealing to 323 K is shown. This spectrum shows a series of vibrational replicas in addition to the main peak corresponding to CHCH (*yellow peaks*). The vibrational splitting was found to be 405 meV. A small amount of CH (*blue peak*) is also detected at this temperature. Reproduced from Ref. [2] with permission from the PCCP Owner Societies

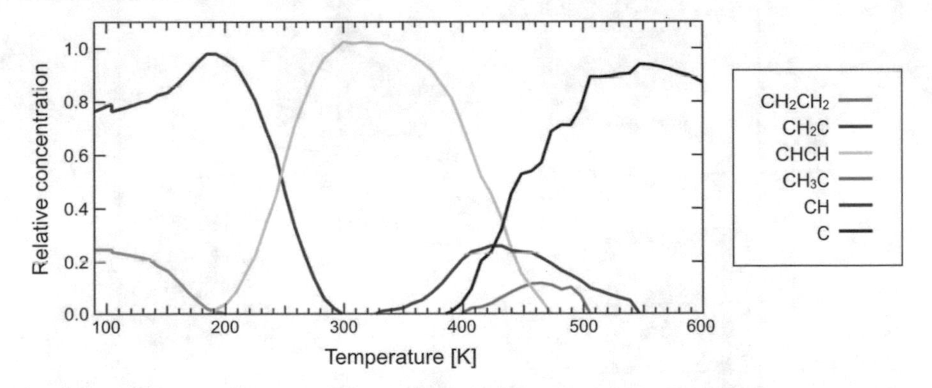

Fig. 3.7 The experimental concentration of the hydrocarbon species on the Ir(111) surface with a temperature ramp of 1.5 Ks^{-1}. The initial coverage (with respect to C monomers) is 0.6 ML. Reproduced from Ref. [2] with permission from the PCCP Owner Societies

The results show that initially both ethylene (C_2H_4) and CH_2C are present on the surface. This, as mentioned before is due to the dehydrogenation of ethylene to CH_2C which occurs almost instantly. At 200 K CH_2C begins to convert to CHCH. Above 300 K CH molecules are formed, most likely from breaking the C-C bond in CHCH. Some CH_3C is formed at around 400 K, along with C monomers. Above 550 K C monomers are the only species present on the surface.

3.6 Reaction Energy Barriers

The first step in determining the correct reaction mechanism from the reaction scheme in Fig. 3.1 is to calculate the energy barriers associated with each reaction. NEB calculations are performed for each reaction in order to determine the minimum energy pathway between the initial state (reactants) and the final state (products). Based on the optimised geometry configurations in Fig. 3.2, the initial and final states for each reaction are constructed and then relaxed to ensure the lowest energy geometry state is found. For the dehydrogenation and hydrogenation reactions the H atom is placed at its most stable top position, some distance from the remaining molecule to minimise the interaction. Likewise for the CC-breaking reactions the CH_m fragments are well separated in the final state.

It is important to note that the direction that the H atom is removed or added can significantly affect the energy barrier for the reaction. For example for a dehydro-

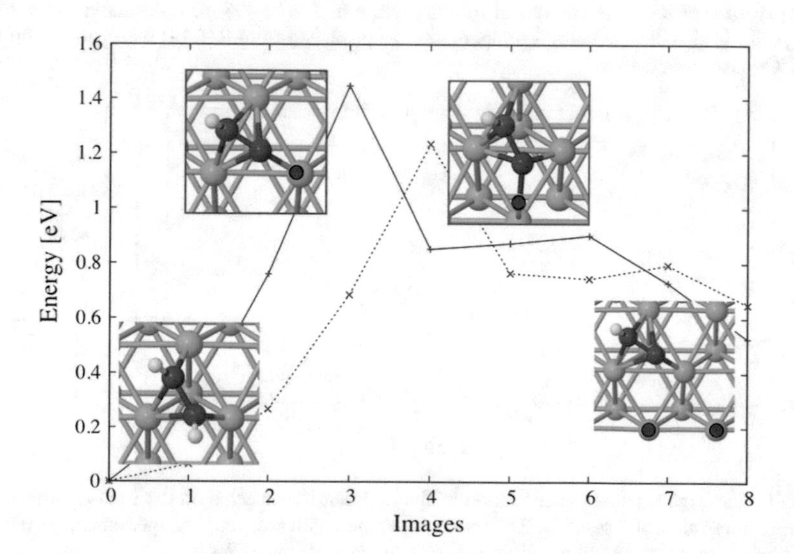

Fig. 3.8 Reaction 13 in detail. The choice of position of the H atom in the final state CHC + H, alters the pathway including the transition state. This changes the energy barrier for the reaction. When placing the H atom in the position marked in *blue* the energy barrier is 0.22 eV lower than when the position marked in *red* is used

Table 3.2 The energy barriers for the various reactions numbered as in Fig. 3.1. E_f is the energy barrier associated with the forward direction of the arrow, whereas E_b is the backward reaction barrier. For reaction CB6 the energy barrier is given as a range as is explained in the Appendix A. Details of the individual calculations (transition states, energy profiles etc.) are given in the Appendix A. Reproduced from Ref. [2] with permission from the PCCP Owner Societies

	Reaction in forward direction	E_f [eV]	E_b [eV]
1	$CH_2CH_2 \rightarrow CH_2CH + H$	0.39	0.58
2	$CH_2CH_2 + H \rightarrow CH_3CH_2$	0.70	0.35
3	$CH_2CH \rightarrow CH_2C + H$	0.35	0.66
4	$CH_2CH + H \rightarrow CH_3CH$	0.64	0.27
5	$CH_3CH_2 \rightarrow CH_3CH + H$	0.33	0.49
6	$CH_3CH_2 + H \rightarrow CH_3CH_3$	2.16	–
7	$CH_2C + H \rightarrow CH_3C$	0.82	0.99
8	$CH_3CH \rightarrow CH_3C + H$	0.46	1.48
9	$CH_2CH_2 \rightarrow CH_3CH$	1.39	1.20
10	$CH_2CH \rightarrow CH_3C$	1.37	2.01
11	$CHC + H \rightarrow CH_2C$	0.54	1.17
12	$CHC \rightarrow C\text{-}C + H$	1.23	0.65
13	$CHCH \rightarrow CHC + H$	1.23	0.58
14	$CHCH + H \rightarrow CH_2CH$	0.77	0.53
15	$CHCH \rightarrow CH_2C$	2.44	2.52
16	$CH_3 \rightarrow CH_2 + H$	0.50	0.58
17	$CH_2 \rightarrow CH + H$	0.09	0.83
18	$CH \rightarrow C + H$	1.11	0.66
CB1	$CH_2CH_2 \rightarrow CH_2 + CH_2$	1.45	0.61
CB2	$CH_3CH_2 \rightarrow CH_3 + CH_2$	1.56	1.47
CB3	$CH_3CH \rightarrow CH_3 + CH$	0.89	1.31
CB4	$CH_2CH \rightarrow CH_2 + CH$	1.07	1.44
CB5	$CH_2C \rightarrow CH_2 + C$	1.88	0.80
CB6	$CHCH \rightarrow CH + CH$	0.79–0.93	1.15
CB7	$CHC \rightarrow CH + C$	0.73	1.29
CB8	$CC \rightarrow C + C$	1.18	1.29
H	$H + H \rightarrow H_2$	1.25	–

genation reaction the position of the hydrogen atom in the final (dehydrogenated) state with respect to its position in the initial intact molecule can affect the energy of the saddle point even though the final state geometries are effectively the same. For this reason the geometries need to be chosen carefully to ensure that the lowest energy pathway for a reaction is found. For some reactions different arrangements of the H atom with respect to the molecule are tested in order to find the lowest possible energy barrier for that reaction. This is illustrated in Fig. 3.8 for reaction 13, CHCH $\rightarrow$ CHC + H.

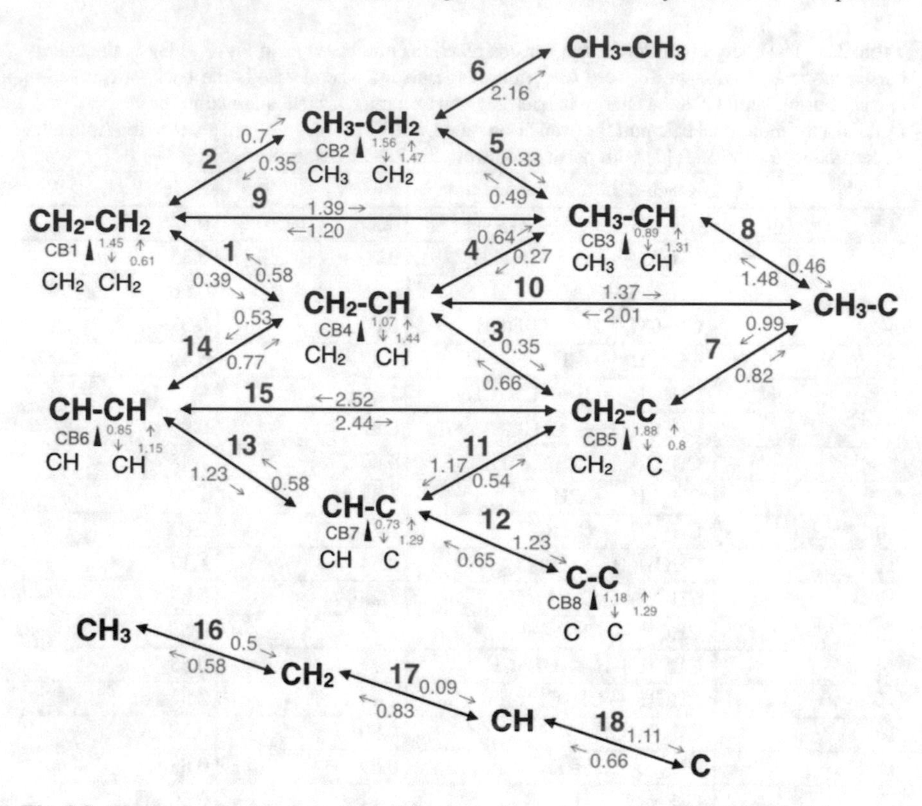

Fig. 3.9 The energy barriers from Table 3.2 arranged onto the reaction scheme in Fig. 3.1

For each reaction in the scheme, the chemical equations and the calculated (lowest energy) forward and reverse energy barriers are shown in Table 3.2. In the Appendix A the initial, final and transition state geometries, and the corresponding energy profiles are presented.

The energy barriers for all the reactions lie in the range 0.3–2.5 eV. The isomerisation reactions, reactions 9, 10, and 15 have much higher barriers than the dehydrogenation and hydrogenation reactions, and will therefore be much less likely to occur. Generally, dehydrogenation reactions have lower barriers than hydrogenation reactions in most cases, apart from reactions 7, 11, 12, and 13. This trend might suggest that dehydrogenation reactions are favoured, eventually leading to the formation of pure carbon species C_n.

Our results also show that the barriers for C-C breaking are mostly of the order of 1–2 eV and are therefore quite high. However, for some of the species involved the barriers may be lower compared to the other possible processes and therefore they must be included in the reaction scheme. Breaking the C-C bond will inevitably lead to CH_m species and then to C monomers. Some of these may also become dehydrogenated or hydrogenated and therefore the barriers for these processes were also calculated. In addition, two single CH_m carbon species can join together to reform the C-C bond, which can be described by the reverse of C-C breaking.

By observation of the energy barrier values and the reaction scheme it may be possible to predict the approximate reaction pathway. Figure 3.9 shows the reaction scheme with the energy barriers placed onto their respective reaction arrows.

By comparing these values with the experimental results it is possible to deduce the reaction pathway leading to the formation of C monomers. Since in the experimental results ethylene dissociates to CH_2C initially, it is most likely that ethylene dehydrogenates twice successively. This is confirmed by the energy barriers which indicate that the lowest energy pathway starting from ethylene is dehydrogenation to CH_2CH, and then to CH_2C. As the temperature increases above 200 K CHCH appears next in the experimental results. This is matched by a decrease in concentration of CH_2C. Judging from the energy barriers the most favourable pathway from CH_2C to CHCH is via hydrogenation to CH_2CH and then dehydrogenation to CHCH. At around 325 K CH is then formed. This must be due to the breaking of the C-C bond in CHCH. In fact the energy barrier for this reaction is around 0.85 eV, and should be possible to overcome at these temperatures. Finally CH dehydrogenates to form C monomers. Since the barrier for hydrogenation of C back to CH is lower than the barrier for CH dehydrogenation this process must be dependent on availability of H atoms on the surface. This means that H atoms have to be lost via the formation of gaseous H_2 molecules in order for many C monomers to be produced.

3.7 Rate Equations

Based on the energy barriers alone without any experimental insight it is difficult to determine how the species will evolve. This is because the chances of different reactions occurring will be dependent on the relative concentrations of the different reacting species. In particular the probability for a hydrogenation reaction to occur will be strongly affected by the availability of H atoms. Since hydrogen may be lost by the formation of H_2, this will limit hydrogenation reactions. The overall reaction process is also made more complicated by the fact that the temperature is being continually increased, such that higher energy barriers will become more easy to overcome with time. In addition there are competing processes with similar barriers.

Hence, in order to fully determine the thermal evolution of the hydrocarbon species on the surface with theory alone the kinetics need to be simulated. A simple way to do this is to solve simultaneous rate equations for the concentration changes in each species. This method is based only on the relative concentrations of the species, and contains no spatial information. To this end, rate equations can only approximate the kinetics at low coverages, assuming the diffusion of all species is fast such that it does not limit any of the reactions.

For each species a rate equation describing its rate of change in concentration is formulated based on the reactions it can undergo as shown in Fig. 3.1. An example for CH_2CH is shown in Fig. 3.10.

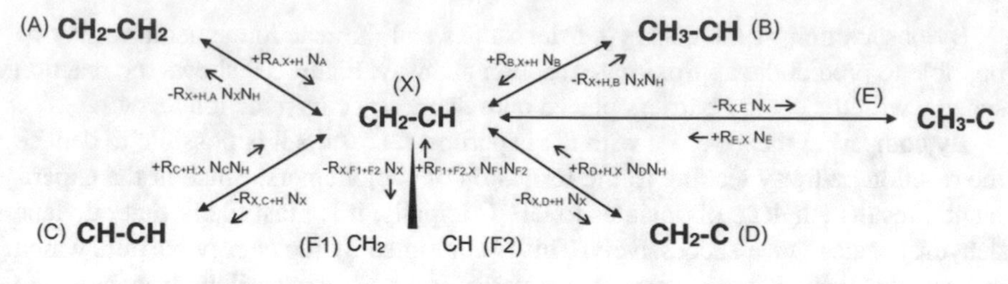

$$\frac{dN_x}{dt} = R_{A,X+H}N_A - R_{X+H,A}N_XN_H + R_{B,X+H}N_B - R_{x+H,B}N_XN_H + R_{C+H,X}N_CN_H - R_{X,C+H}N_X$$
$$+R_{D+H,X}N_DN_H - R_{X,D+H}N_X + R_{E,X}N_E - R_{X,E}N_X + R_{F1+F2,X}N_{F1}N_{F2} - R_{X,F1+F2}N_X$$

Fig. 3.10 The rate equation for the change in concentration of CH_2CH (X) in terms of its possible reactions as adapted from Fig. 3.1. The rate for a process $i \rightarrow j$ is given as $R_{i,j}$ and N_i denotes the concentration of a species i. N_H refers to the concentration of hydrogen for the hydrogenation reactions. All other species are labeled with letters

Each reaction contributes two terms, one with a positive sign and the other with a negative sign, describing the increase and decrease in the concentration of the species in question due to the forward and reverse reaction processes. Each term for a reaction between states $i \rightarrow j$ is comprised of the reaction rate, given in the usual Arrhenius form, $R_{i,j} = \nu_{i,j} \exp(-E_{i,j}/k_BT)$, and the concentrations of the relevant reactant species. The reaction terms can be either unimolecular (A → B + C) or bimolecular (B + C → A). Unimolecular terms have the form $N_A R_{A,B+C}$, whereas bimolecular terms have the form $N_B N_C R_{B+C,A}$. To calculate the rates, both the reaction energy barriers and the pre-exponential factors are required. The barriers are supplied from the NEB results in Table 3.2. The pre-exponential factors can be calculated using the Vineyard formula (Eq. 2.43), however in most cases it is suitable to take $\nu_{i,j} = 10^{13}$ s^{-1}. For the reactions which are crucial to the reaction pathway (as indicated from initial results) the pre-exponential factors and the energy barriers are calculated with high accuracy, requiring up to six layers of iridium to model the surface. More details of this are given in the Appendix A.

The rate equations are solved simultaneously using the Runge-Kutta ordinary differential equation method. The was implemented in a code by J. Posthuma de Boer. The rate equations for each species are contained in the Appendix A. The simulation is performed to imitate the experimental conditions. It is initiated with an initial concentration of ethylene molecules in ML. Throughout the simulation the temperature is increased linearly from 100 K with a rate of 1.5 Ks^{-1}, as in the experiment.

In Fig. 3.11 the thermal evolution of species as determined from the rate equations is shown for an initial ethylene coverage of 0.6 ML. At the start of the simulation ethylene (C_2H_4) converts rapidly to CH_2C. At around 250 K both CHCH and CH_3C appear whilst the concentration of CH_2C decreases. CHCH is short-lived since it can readily break into two CH molecules (blue curve). However CH_3C survives until about 450 K when the temperature becomes large enough so that it can convert to CH. This presumably occurs via the reaction pathway $CH_3C \rightarrow CH_2C \rightarrow CH_2CH$

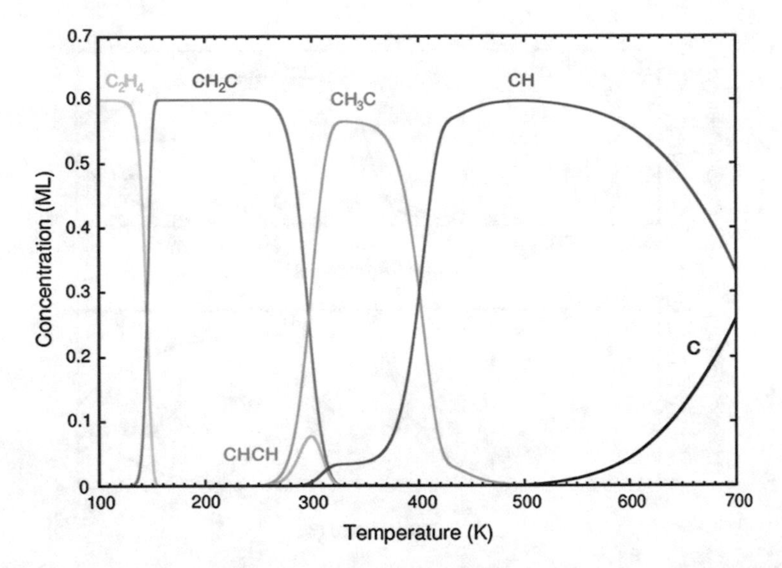

Fig. 3.11 The concentration of the hydrocarbon species on the Ir(111) surface as determined from solving the rate equations. As with the experimental results, there is a temperature ramp of 1.5 Ks^{-1} and the initial coverage of ethylene is 0.6 ML

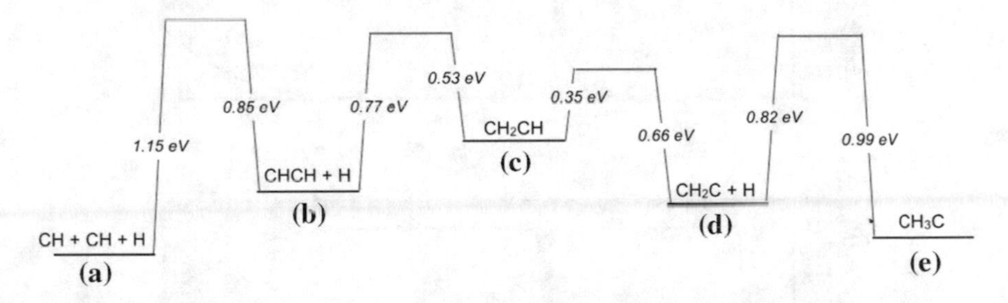

Fig. 3.12 The energy profile of the various important competing reaction processes, illustrating the differences in barrier heights

→ CHCH → CH, as indicated by the height of the energy barriers. CH remains until it can dehydrogenate to C. Since the barrier for this is larger than the reverse hydrogenation reaction this can only occur once enough H has been lost via the formation of H_2 to prevent CH from reforming.

The evolution of species, as determined by the rate equations, can be explained in terms of the energy barrier profile for the reactions. The energy profile for reactions involving the intermediate species, CH_2CH, CH_2C, CH_3C, CHCH and CH, is shown in Fig. 3.12. Ethylene, which is the initial species dehydrogenates to CH_2CH, since the barrier for it to do so is sufficiently low enough. From this point onwards, CH_2CH, which is labeled as (c) in Fig. 3.12, may easily dehydrogenate to CH_2C + H (d) since the barrier is only 0.35 eV. CH_2C is stable on the surface until the temperature is increased sufficiently. It may hydrogenate to CH_2CH or CH_3C. The latter has a

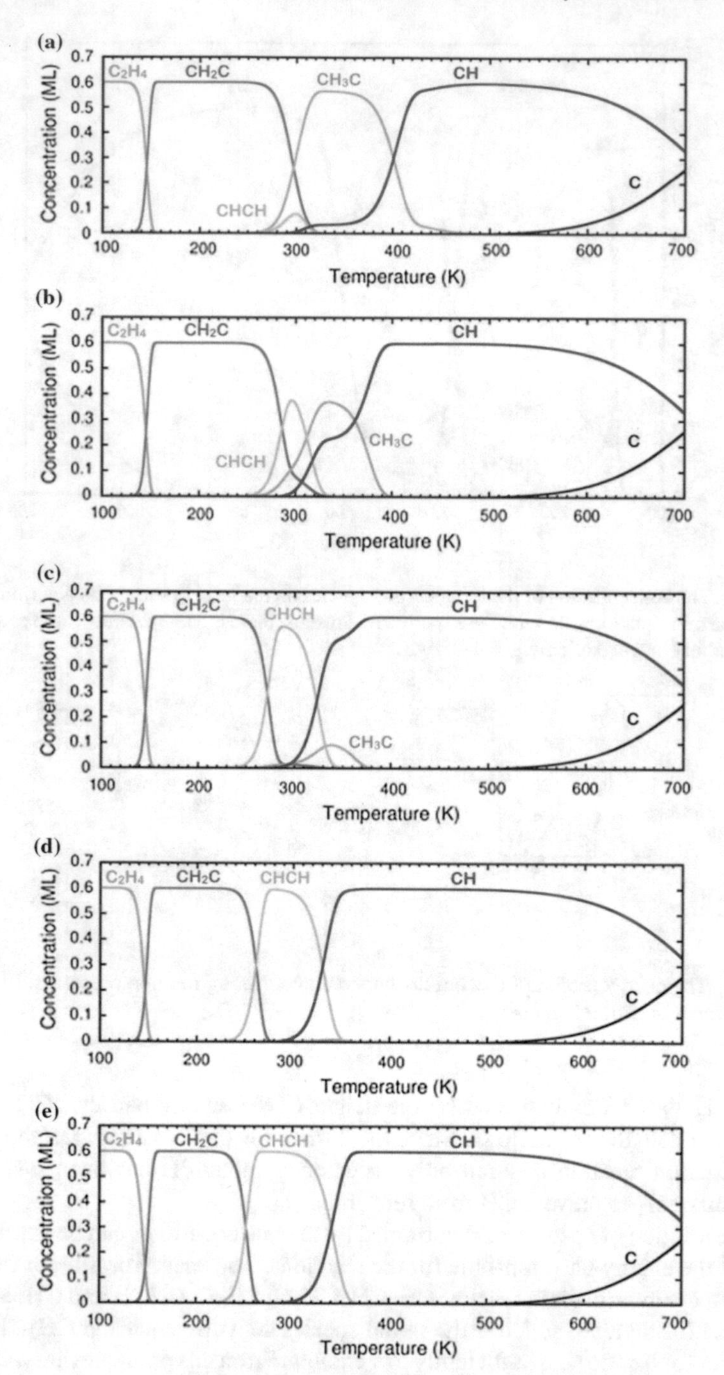

Fig. 3.13 The thermal evolution of species on the surface as calculated from solving the rate equations. The pre-exponential factors are scaled between their calculated value **a** and the altered value **e** as displayed in Table 3.3. The pre-factors in **b**, **c** and **d** are scaled by 25, 50 and 75% respectively

Table 3.3 The calculated and scaled pre-factors ν

Reaction	ν^{calc} [s^{-1}]	$\nu^{adjusted}$ [s^{-1}]
$CH_2CH \rightarrow CH_2C + H$	1.7×10^{12}	1.7×10^{11}
$CH_2C + H \rightarrow CH_2CH$	5.2×10^{12}	5.2×10^{13}
$CH_2C + H \rightarrow CH_3C$	8.9×10^{12}	8.9×10^{11}
$CH_3C \rightarrow CH_2C + H$	1.9×10^{12}	1.9×10^{13}
$CH_2CH \rightarrow CHCH + H$	3.1×10^{12}	3.1×10^{13}
$CHCH + H \rightarrow CH_2CH$	1.6×10^{12}	1.6×10^{11}
$CHCH \rightarrow CH + CH$	3.6×10^{12}	3.6×10^{11}
$CH + CH \rightarrow CHCH$	5.9×10^{11}	3.6×10^{11}

higher energy barrier, however hydrogenation to CH_2CH will be short lived, since dehydrogenation is more favourable. CHCH (b) may be formed if the 0.53 eV barrier is overcome, however since CH_2CH is unstable there is a short time frame for this to occur. Meanwhile if CH_2C hydrogenates to CH_3C (e), CH_3C will remain for a long time, since the 0.99 eV barrier traps it in this state. It is only once the temperature has risen to ~350 K that the more stable state, when CH is formed (a), can be accessed. The energetics of these reactions results in the thermal evolution of species as shown in Fig. 3.11.

Compared to the experimental results (Fig. 3.7) the results from the rate equations above contain some similarities and some notable differences. The overall reaction pathway from ethylene to C monomers appears similar in both cases since many of the intermediate species are the same. However the experimental results only predict a small amount of CH_3C molecules formed during the decomposition between 300 and 400 K, while CHCH is formed in large amounts after CH_2C and exists for much longer (until around 460 K). This is contrary to the rate equations results. These differences indicate that solving the rate equations to determine the species evolution does not give a good approximation to the actual experimental coverages, especially since high coverages are considered, where spatial information becomes important. This results in CH_3C and CH being produced too easily compared to the actual decomposition process, which in turn reduces the lifetime of CHCH.

Spatial effects, such as the coverage and diffusion of species can be manifested by effectively scaling the rates of different reactions. This can be included as an additional scaling constant to the pre-exponential factors to the reaction rates. By comparing with the experimental results the effect of the spatial effects to the individual reaction rates can be estimated. This is demonstrated below by altering the pre-exponential factors of various reactions in order to show how the calculated coverages are easily changed to become more similar to the experimental results. The pre-exponential factors are adjusted by one order of magnitude (up or down) which could easily be accounted for by spatial effects. Figures 3.13a–e show that over this range of pre-factors the concentration of CH_3C can be reduced to zero and the concentration of CHCH can be increased.

Even with the pre-factor adjustment the coverages are not exactly comparable to the experimental results. This is because the pre-factors are adjusted to a fixed value. In fact the spatial effects can depend on the temperature and other factors such as the surface coverage, and hence the reaction rates will vary over the course of the decomposition process. In order to accurately simulate these effects a kinetic Monte Carlo scheme can be devised. This will be the subject of Chap. 4.

3.8 Conclusions

In this chapter the thermal decomposition mechanism for ethylene on the Ir(111) surface was investigated using both experimental and theoretical techniques. By considering the relevant hydrocarbon species and then constructing a reaction scheme, as shown in Fig. 3.1, the energy barriers for all possible reactions were calculated. From these barriers the kinetics of a system under a constant temperature ramp was determined by solving a system of rate equations. This deduced the evolution of species on the surface as shown in Fig. 3.11. Based on these results it was found that the decomposition mechanism starts with the dehydrogenation of ethylene to CH_2C. After this some CH_2C is converted to CHCH molecules, however most is hydrogenated to CH_3C. Once the temperature is above 350 K the barrier to dehydrogenate CH_3C is overcome and CH is formed via breaking of the C-C bond in CHCH. Finally CH dehydrogenates to C monomers.

The experimental species coverages were found using *in situ* XPS measurements. The spectra corresponding to the observed species were identified by calculating the core level binding energies for the C 1s level of each species. The experimental coverages are shown in Fig. 3.7. Qualitatively the results are similar to the theoretical results, except for the relevant amounts of the intermediate species CH_3C and CHCH. The experimental results predict that CHCH should be the dominant species over the temperature range of 200–450 K. Instead the theoretical results find only a small amount of CHCH is formed and that CH_3C is prevalent over this temperature range.

Based on both the experimental coverages and the predominant reaction mechanisms determined from the rate equations, the following reaction pathway for the decomposition of ethylene on the Ir(111) surface can be determined.

$$CH_2CH_2 \rightarrow CH_2CH + H$$
$$CH_2CH \longleftrightarrow CH_2C + H$$
$$CH_2C + H \longleftrightarrow CH_3C$$
$$CH_2CH \rightarrow CHCH + H$$
$$CHCH \rightarrow CH + CH$$
$$CH \rightarrow C + H$$

This reaction pathway shows the importance of all reaction types in the decomposition. Both hydrogenation and dehydrogenation reactions are required to form CHCH. The breaking of the C-C bond, which occurs once CHCH has been formed allows single C species to be produced. After this, the CH molecules dehydrogenate to C monomers. These may then go on to become the building blocks for graphene growth. An important feature of the pathway is that C monomers are formed rather than dimers. This allows the possibility for graphene to be built from clusters which contain an odd number of C atoms when ethylene is used as a precursor, as was suggested in [4, 5].

Differences between the experimental and theoretical results may be due to the lack of spatial information included in the rate equations. By increasing or decreasing the pre-exponential factors by one order of magnitude the calculated evolution of species changes drastically and can be made to fit more closely with the experimental results as demonstrated in Fig. 3.13. It is worth noting that the energy barriers themselves still contain a margin of error which could affect the rate equation results.

References

1. H.A. Aleksandrov, L.V. Moskaleva, Z.-J. Zhao, D. Basaran, Z.-X. Chen, D. Mei, N. Rösch, Ethylene conversion to ethylidyne on Pd(111) and Pt(111): A first-principles-based kinetic Monte Carlo study. J. Catal. **285**(1), 187–195 (2012)
2. H. Tetlow, J. Posthuma de Boer, I.J. Ford, D.D. Vvedensky, D. Curcio, L. Omiciuolo, S. Lizzit, A. Baraldi, L. Kantorovich, Ethylene decomposition on Ir(111): initial path to graphene formation. Phys. Chem. Chem. Phys. **18**, 27897–27909 (2016)
3. W.P. Krekelberg, J. Greeley, M. Mavrikakis, Atomic and molecular adsorption on Ir(111). J. Phys. Chem. B **108**(3), 987–994 (2004)
4. E. Loginova, N.C. Bartelt, P.J. Feibelman, K.F. McCarty, Evidence for growth by C cluster attachment. New J. Phys. **10**(093026) (2008)
5. E. Loginova, N.C. Bartelt, P.J. Feibelman, K.F. McCarty, Factors influencing graphene growth on metal surfaces. New J. Phys. **11**(063046) (2009)

Chapter 4
Hydrocarbon Decomposition: Kinetic Monte Carlo Algorithm

This section follows on from the previous section where the decomposition of ethylene on the Ir(111) surface was studied using DFT methods and rate equations. After the calculation of the relevant reaction energy barriers with the NEB method rate equations were solved in order to determine the thermal evolution of the hydrocarbon species on the surface. However rate equations contain no spatial information about the decomposition process and can therefore only give an approximate simulation of the occurring processes. In this chapter a lattice-based kinetic Monte Carlo simulation is devised to model the ethylene decomposition process.

4.1 Method

The kMC code is required to simulate the ethylene decomposition process, based on the reaction scheme in Fig. 3.1. By modelling the species on a lattice, which represents the substrate, spatial information can be included from knowledge about the positions of the species on the lattice grid. At each time step a list of possible reactions is generated based on the current configuration and then from this a reaction is selected with a probability proportional to its rate. The difficultly in the algorithm is determining which reactions are possible at each given time step. This is complicated by the large number of reactions, each of which will only be possible under a certain set of circumstances, depending on the species and their local environment.

The main ingredients of the kMC algorithm are the various object lists containing the surface grid sites, the current species, and the possible reactions for the next step (and the corresponding possible new species).

The species list contains information about the current species on the surface such as their type, their orientation on the surface and the sites that are occupied by them. The surface sites list contains information about whether sites are occupied or unoccupied, and if occupied which species type is occupying them. This is so that the occupation of sites located near a particular species can be easily checked during the reaction finding process.

H.A. Tetlow, *Theoretical Modeling of Epitaxial Graphene Growth on the Ir(111) Surface*, Springer Theses, DOI 10.1007/978-3-319-65972-5_4

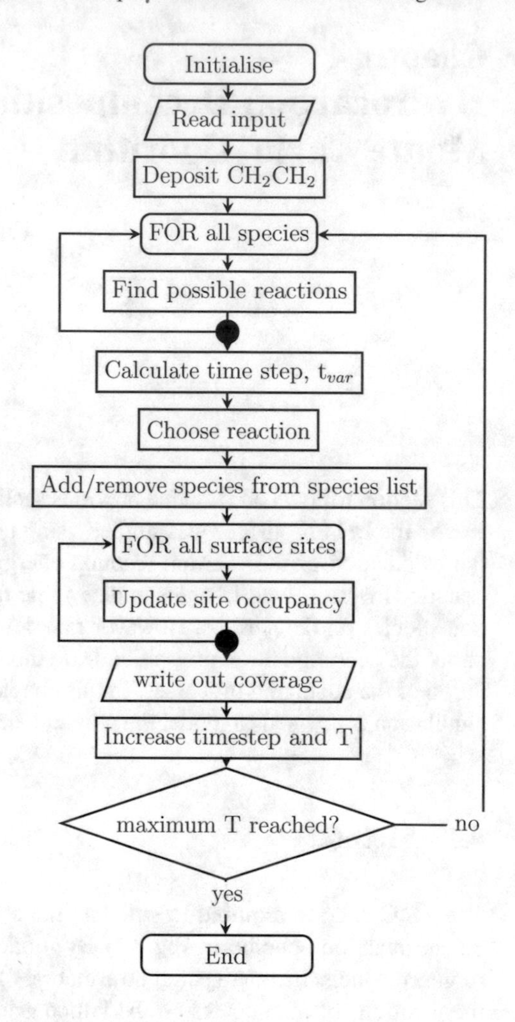

Fig. 4.1 The main algorithm for the kMC code to simulate ethylene decomposition

The general algorithm for the kMC code is shown in Fig. 4.1. Initially a given number of ethylene (CH_2CH_2) molecules are placed randomly onto the surface grid, with random orientations.

During one reaction finding step the code examines each species on the surface. For each species different types of reaction (hydrogenation/dehydrogenation, isomerisation, C-C breaking/recombination and H_2 formation) are searched for based on:

- The occupation of neighbouring top sites: whether these are occupied by H, or unoccupied (for hydrogenation or dehydrogenation reactions).
- The occupation of other neighbouring sites: whether these are occupied by CH_m species if the current species is also CH_k (for C-C recombination reactions).

- The fitting on the new species: whether the new species can fit on surface given the location of the current species (for all reactions).

Details of these processes depending on the type of reaction are given in Sect. 4.4. During the reaction finding step the possible reactions, and the new species associated with them are each added to lists. The rates of the new reactions are then summed to give the total rate, which is used to calculate the variable time step, t_{var}. This time step determines how much time will pass before the reaction can occur. The method for calculating it will depend on whether the temperature is fixed or variable.

Diffusion moves can be incorporated together with the reaction moves with the full kMC treatment. However diffusion can also be treated separately for convenience and to speed up to simulation. In this case after the time step is determined diffusion of the species is performed. If the energy barriers for diffusion are found to be on average much smaller than for the reactions many diffusion processes can occur before the next reaction move is made. It is possible, if the diffusion processes are fast, to instead randomly move each of the species on the surface. This reduces the computational effort.

After diffusion has been performed the routine checks for reactions again, since the species will have moved. For these the kMC algorithm is used to determine the new move based on their rates. Once the new move has been chosen the reactant species are removed from the species list, and the new species are added to it. The occupation of the grid sites is updated based on the species list. Once the time step (and if necessary the temperature) have been increased information about the current species are output. This includes the coverage, reaction evolution and a graphical representation of the surface. At the end of the step all the unnecessary data about the possible reactions and new species is cleared. Until the maximum time or temperature is reached the process repeats, starting from determining the new reactions and time step.

4.2 Surface Lattice Grid

The lattice for the fcc(111) surface is built up of interlocking triangles with circles centred on the points where the six neighbouring triangles intersect. An image of the surface grid and the actual surface are shown in Fig. 4.2a and b respectively.

Each triangle represents a hollow site on the surface, whereas each circle represents a top site. In order to translate between neighbouring grid sites mapping functions are used. For a triangle site the grid numbers of the three neighbouring triangle sites and of the three neighbouring circle sites can be easily accessed. Likewise for a circle site the six neighbouring triangle sites can also be found. Periodic boundary conditions are applied.

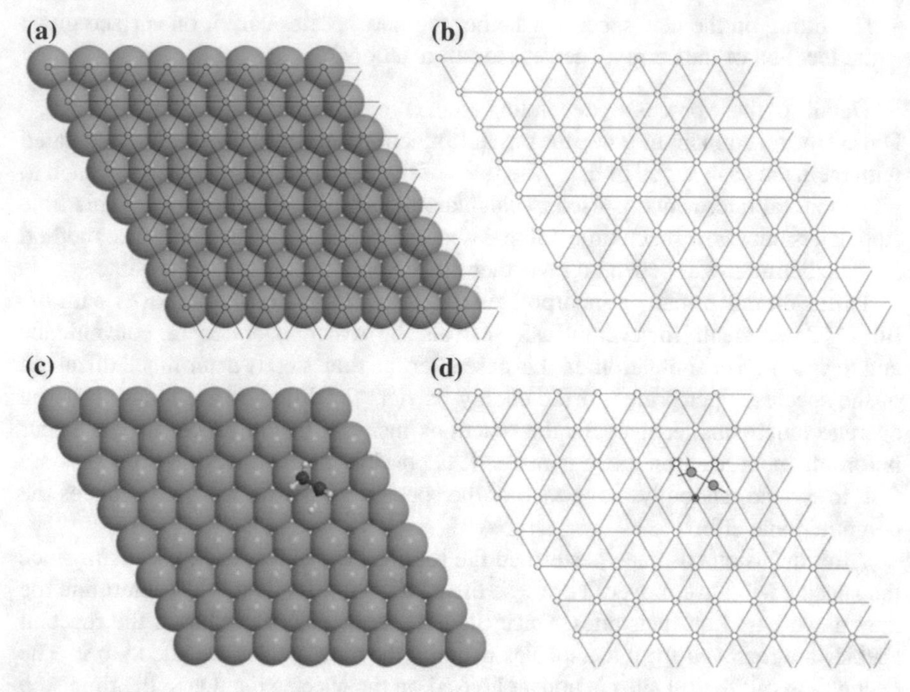

Fig. 4.2 **a** A section of the fcc(111) surface and, **b** the model of the same surface used in the simulation which is comprised of triangular hollow sites and *circle top* sites. **c** A CH_2CH molecule and an H atom adsorbed on the surface occupying two hollow sites and a *top* site and a single *top* site respectively. **d** The cartoon drawing of the configuration in **c** as output from the kMC simulation

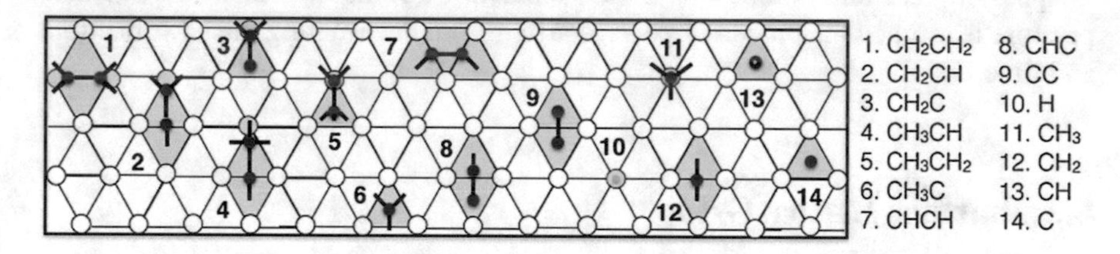

Fig. 4.3 The 14 different species which are included in possible reactions in the KMC simulation. The grid sites shaded in *blue* show the sites that are considered occupied by each of the species and a cartoon figure of the molecule is overlaid on *top*. Each species is given in its 0° orientation

4.3 Hydrocarbon Species

Each hydrocarbon, depending on its shape may occupy up to four neighbouring grid sites, which can be either circle or triangle grids. These are decided by simple observation of the lowest energy geometry configurations. The grid sites associated with each species are shown in Fig. 4.3.

Table 4.1 The rotational degrees of freedom for each of the species in the kMC model. For six degrees of freedom the species can orientate at either 0°, 60°, 120°, 180°, 240° or 300°. For three degrees of freedom the species orientation will be either 0°, 120° or 240°

Species	Degrees of freedom
CH_2CH_2	3
CH_2CH	6
CH_2C	6
CH_3CH	6
CH_3CH_2	6
CH_3C	1
CHCH	6
CHC	6
CC	3
CH_3	1
CH_2	3
CH	1
C	1
H	1

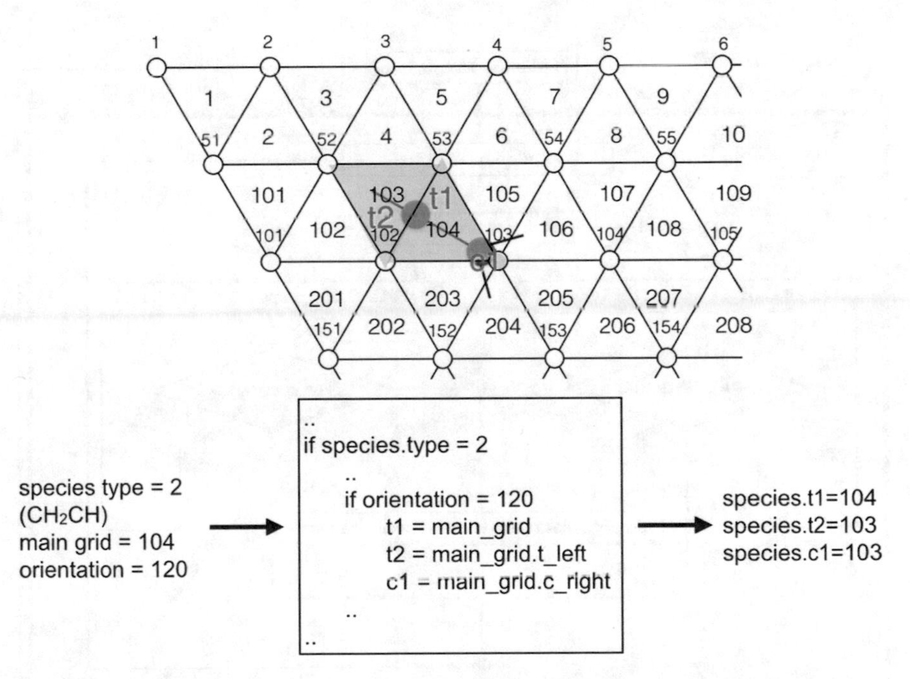

Fig. 4.4 An illustrated example of how the kMC code deduces the sites occupied by a species depending on its type, orientation and "main grid" position

Each species added to the surface is assigned by its "main-grid" position, which is the number of one of the grid sites (triangle or circle) that it occupies. For most of the species different orientations are possible depending on their symmetry with the lattice. The rotational degrees of freedom for each species are given in Table 4.1.

Each species is defined by its type (a number from 1 to14), its main grid site number and its orientation in degrees (0°, 60°, 120°, 180°, 240°, 300°). Based on these three quantities the remaining occupied grid sites are deduced using a function. An example of this is shown in Fig. 4.4 for CH_2CH.

For each species added to the surface the grid sites that it occupies are added to the species list. In turn the occupation of the grid sites is updated to reflect that the occupation of the sites has changed.

4.4 Reactions

At each time step a list of all the possible reactions is generated by inspecting each species individually and then finding the potential reactions based on the occupation of its neighbouring sites. The reaction list contains information about the identity of the reactant and product species, as well as the reaction type and its energy barrier.

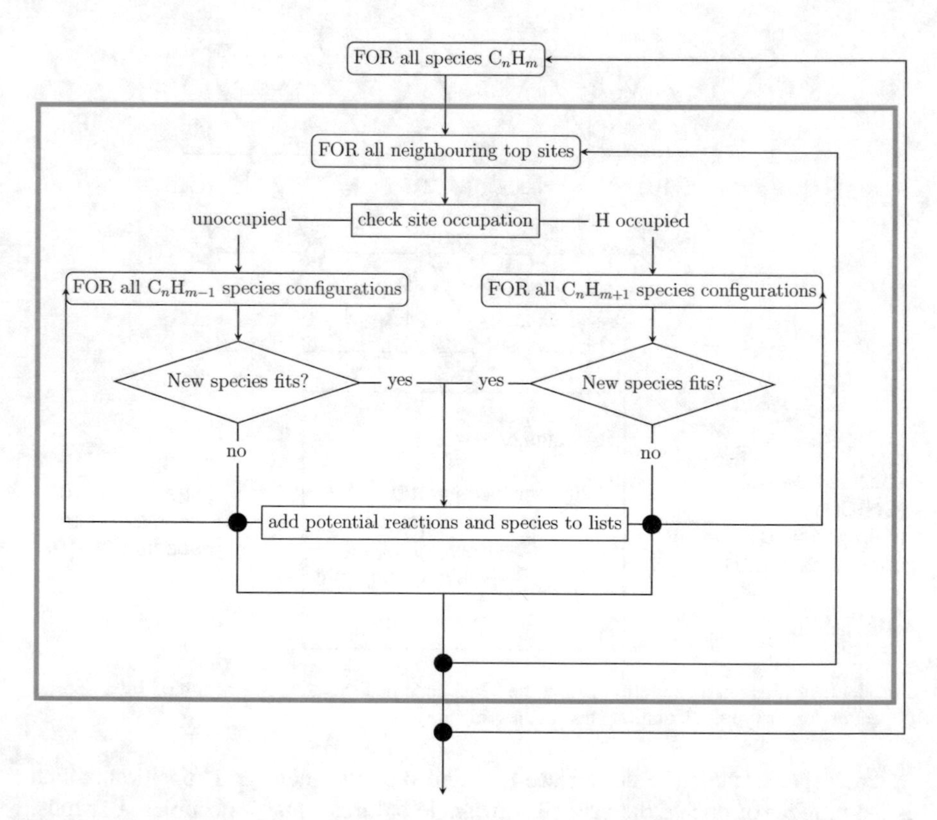

Fig. 4.5 The algorithm for checking for possible hydrogenation and dehydrogenation reactions (inside the *grey box*). *Black dots* indicate where the code exits the FOR loops

Reactions are divided into 7 different types since the type will affect the outcome of the reaction and also the reaction checking method will differ.

4.4.1 Hydrogenation and Dehydrogenation Reactions

The reaction checking process for hydrogenation and dehydrogenation reactions is shown in Fig. 4.5. To check for the reactions of a single C_nH_m species first the occupation of a selection of its neighbouring top sites are checked. These sites are predefined depending on the type of species, and are based on the exact reaction mechanism determined by the NEB calculations.

If any number of the neighbouring top sites are unoccupied then a dehydrogenation reaction may be possible. Depending on the location of the top sites that are unoccupied then the products of the dehydrogenation reaction maybe be different (for some species). For example when removing an H atom from CH_2CH it can be taken from either the CH_2 half or the CH half, resulting in CHCH or CH_2C respectively. For each dehydrogenation reaction the approximation is made that the rate is the same regardless of the final position of thc H atom. The same logic is applied

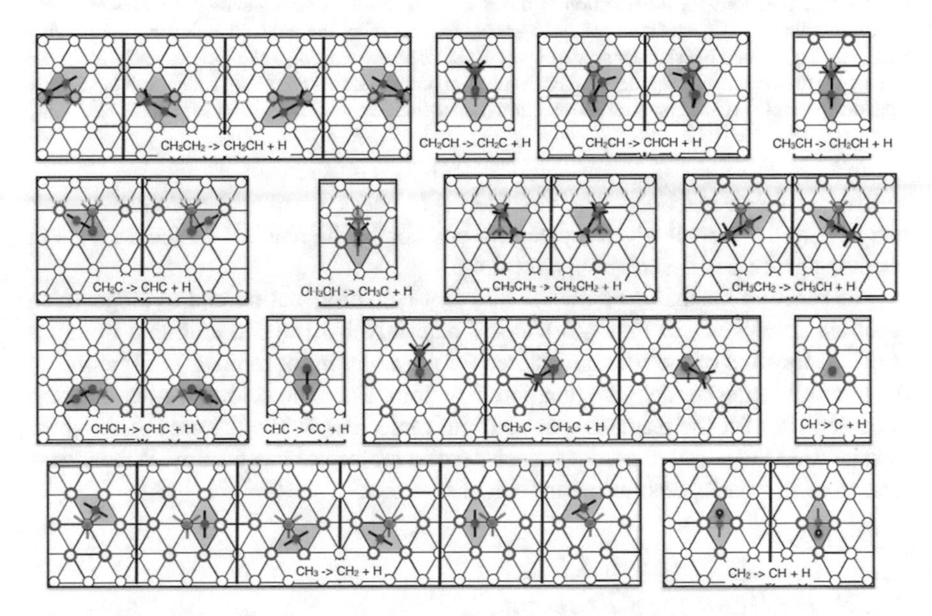

Fig. 4.6 The dehydrogenation reactions for each species $C_nH_m \rightarrow C_nH_{m-1} + H$. The *red circles* show the possible positions for the removal of H atoms. The *blue shading* shows the sites occupied by the reactant, C_nH_m, whereas the light *green shading* show the sites occupied by the product, C_nH_{m-1}. These may overlap as indicating by the *dark green/turquoise shading*. In each image the reactant is orientated at 0 degrees. For each reaction (enclosed in a *box*) the product can have more than one possible orientation after the reaction (hence there are multiple images for some reactions)

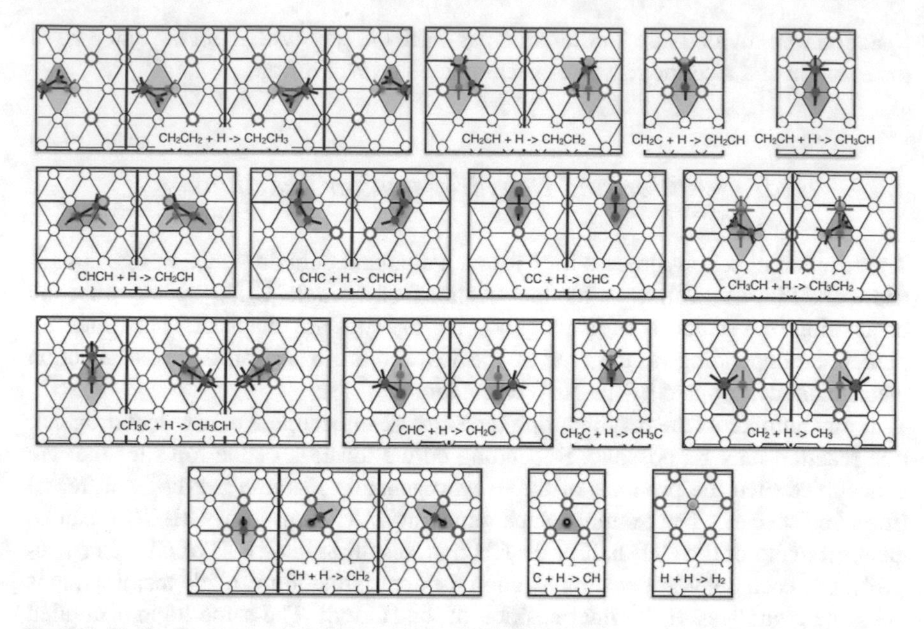

Fig. 4.7 The hydrogenation reactions for each species $C_nH_m + H \rightarrow C_nH_{m+1}$. The *red circles* show the possible positions for the addition of the H atom. The *blue shading* shows the sites occupied by the reactant C_nH_m, whereas the *light green shading* show the sites occupied by the product C_nH_{m+1}. These may overlap as indicating by the *dark green/turquoise shading*. In each image the reactant is orientated at 0 degrees. For each reaction (enclosed in a *box*) the product can have more than one possible orientation after the reaction (hence there are multiple images for some reactions)

to hydrogenation reactions. The next step is to check whether the product species can fit onto the surface given the current position of the reactant molecule. This fit checking process is described in Sect. 4.4.6.

The reaction mechanisms for the different dehydrogenation and hydrogenation reactions are shown in Figs. 4.6 and 4.7 respectively. The red coloured top sites show the location of the sites considered for the addition (or removal) of H atoms in the hydrogenation (or dehydrogenation) reactions. The blue coloured sites show the occupied sites for the reactant species (faint molecule), while the green sites show the sites that the possible product species (solid molecule) can occupy. Where these sites overlap a dark green/turquoise colour is shown.

4.4.2 H_2 Desorption Reaction

Hydrogen atoms may be lost from the surface via the formation of $H_2(g)$. The NEB calculation for this process shows that H_2 is unstable on the surface, and therefore the formation of H_2 and its desorption must happen in a single step. This is detailed in the Appendix A. Therefore the process for determining if a H_2 desorption reaction

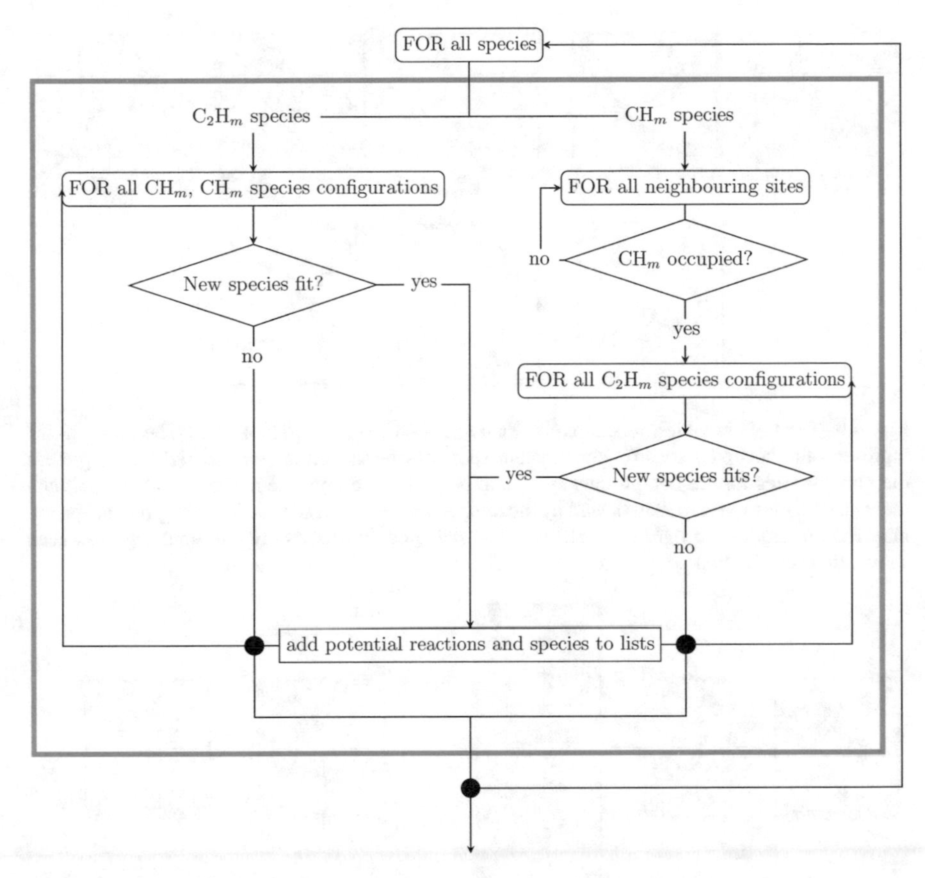

Fig. 4.8 The algorithm for the checking of possible C-C breaking and C-C recombination reactions (inside the *grey box*). *Black dots* indicate where the code exits the FOR loops

can occur is similar to checking for hydrogenation reactions except there is no need to check for fitting of the product species. For each H atom on a top site its six neighbouring top sites are checked for other H atoms. If present then the reaction is added to the possible reactions list. The total rate for these reactions is divided by two since each possible reaction is double counted. If the reaction is chosen then both H atoms are removed from the species list.

4.4.3 C-C Breaking and C-C Recombination Reactions

The algorithm for the determination of the C-C breaking and recombination reactions is shown in Fig. 4.8.

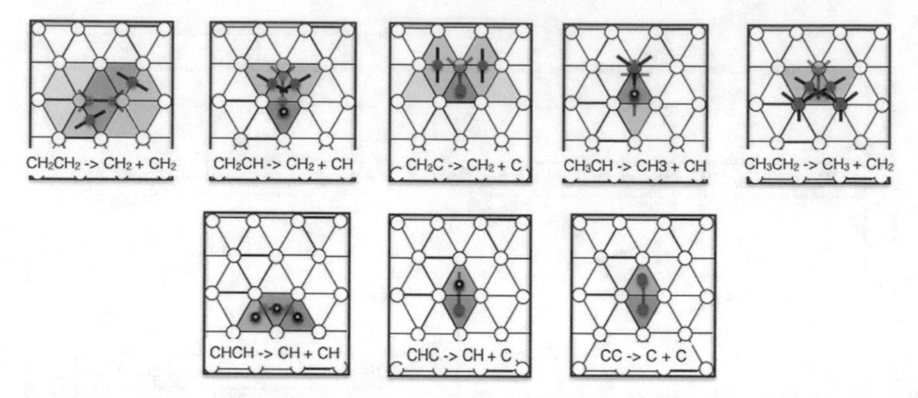

Fig. 4.9 The C-C breaking reactions for each species $C_nH_m \rightarrow CH_k + CH_l$. The *blue shaded* represents the occupied sites for the reactant species, whereas the *green* and *red* sites represent the possible sites for the two product species to be positioned after the reaction. The product and reactant sites may overlap as indicated by the *dark green* and *dark red shading*. The product species may form in any of the *lightly shaded green* or *red* sites, however only one final state has been shown for each reaction

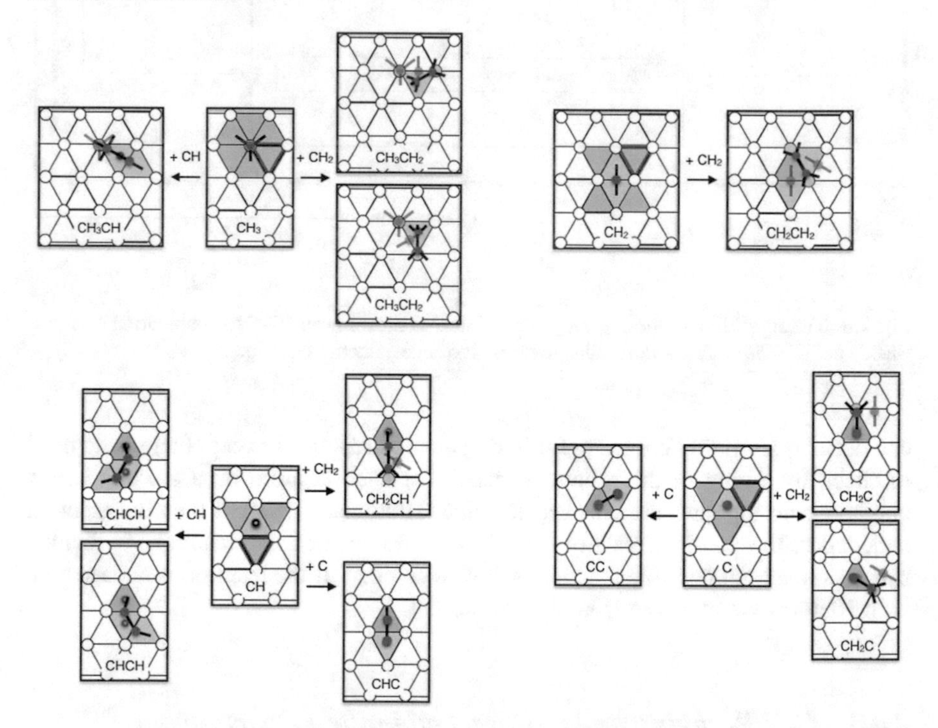

Fig. 4.10 The C-C recombination reactions for each species $CH_k + CH_l \rightarrow C_nH_m$. For the CH_3, CH_2, CH and C species the neighbouring sites (*shaded red*) are checked for other CH_m species. If one of these is present (for example in the site with the *red* outline) then the product, C_nH_m species (which depends on the CH_k, CH_l species) can be formed as shown in the *green shaded* sites. In some case there are two possible orientations for the product species, both of which are shown

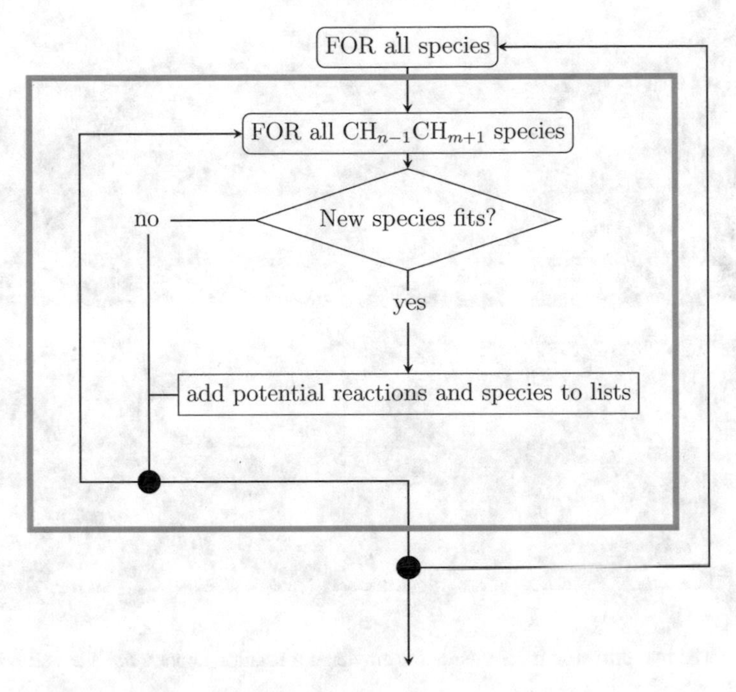

Fig. 4.11 The algorithm for the checking of possible isomerisation reactions (inside the *grey box*). *Black dots* indicate where the code exits the FOR loops

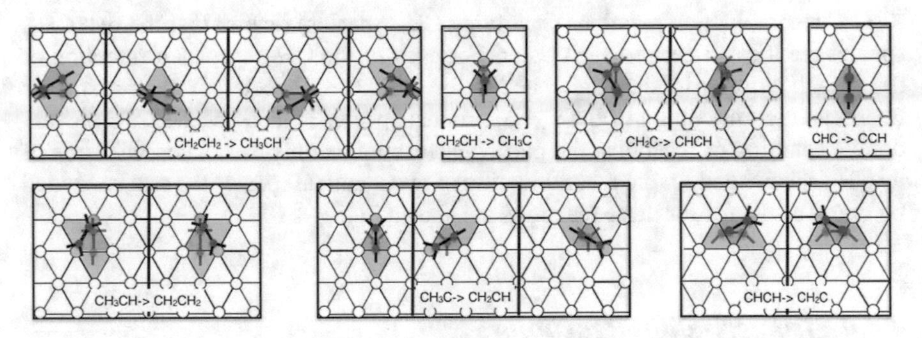

Fig. 4.12 The isomerisation reactions for species A → B. The sites occupied by the reactants A are *shaded blue*, whereas the sites occupied by the products B are *shaded green*. The sites where the product and reactant overlap are *shaded dark green/turquoise*

For C C breaking reactions (involving C_2H_m species) the two produced species must be arranged on the surface in the available sites. For many of the species there are multiple possible ways of achieving this. These are shown in Fig. 4.9. If both of the species can fit on the surface in any of the combinations in Fig. 4.9 then the reaction is possible and both the new fragments are added to the potential species list.

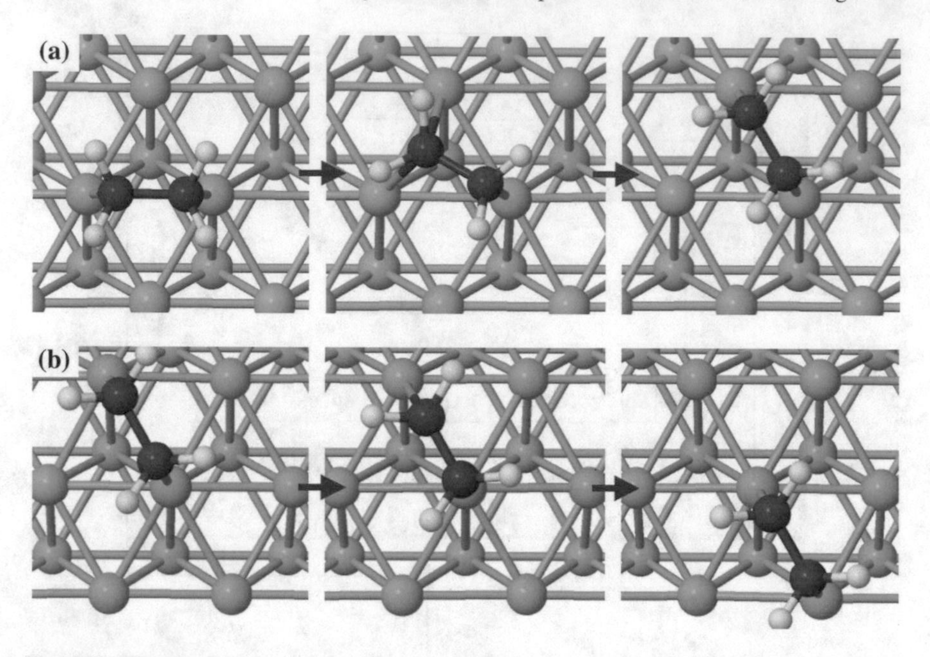

Fig. 4.13 The two diffusion mechanisms for ethylene: **a** rotation about a *top* site and **b** hopping over a *top* site

C-C recombination reactions (involving CH_m species) require that the two CH_m species are located near to each other. For each of the CH_m species a selection of the neighbouring grid sites are checked for other CH_m species. If such a species is present then the fitting of the C_nH_m product species is then checked. To prevent double counting of reactions, the particular reaction is only associated with one of its CH_m species and for the recombination of two identical species the rate is divided by two. The details for these reactions are shown in Fig. 4.10.

4.4.4 *Isomerisation Reactions*

Isomerisation reactions only involve one species molecule, so no scanning of neighbouring surface sites is necessary before checking the fitting of the product species. If a particular hydrocarbon can isomerise into another then the fitting of the new molecule is checked. The reaction checking algorithm is shown in Fig. 4.11 and details of the isomerisation reactions are given in Fig. 4.12.

4.4.5 Diffusion

Compared to the rate equation approach discussed in the previous chapter, one of the main difference with kMC simulations is that spatial effects are included. Since molecules are considered as separate entities on the surface they must be allowed to diffuse around on the surface in order to come into contact with each other and react. In order to determine the mobility of the different species the diffusion barriers must first be calculated. To do this two different diffusion mechanisms are considered: rotation and hopping. For rotation one half of the molecule remains in the same position, while the other half rotates around it. This diffusion mechanism can only be applied to the C_2H_m species since they (usually) have two points of contact with the surface. In hopping diffusion the molecule "slides" along the surface. An example of both of these diffusion mechanisms is shown in Fig. 4.13 in the case of ethylene. Figures showing the exact diffusion mechanisms are found in the Appendix. The energy barriers, as determined by NEB calculations, for the diffusion mechanisms of all the different hydrocarbon species are shown in Table 4.2.

Table 4.2 The energy barriers for diffusion of the species on the surface shown together with the corresponding type of diffusion mechanism. For species moving between hollow sites the barrier may be affected by whether the site is fcc or hcp. In this case we take the highest energy barrier since in order for a molecule to diffuse freely on the surface it must pass across both types of hollow site

Species	Diffusion mechanism	E_{diff} [eV]
CH_2CH_2	Rotation	0.40
CH_2CH_2	Hopping	0.37
CH_2CH	Hopping	0.51
CH_2CH	Rotation	0.51
CH_2C	Hopping	0.99
CH_2C	Rotation	0.99
CH_3CH	Hopping	0.33
CH_3CH_2	Hopping	0.73
CH_3C	Hopping	0.51
CHCH	Rotation	0.33
CHCH	Hopping	0.56
CHC	Rotation	0.16
CHC	Hopping	0.53
CC	Rotation	0.90
CC	Hopping	0.83
CH_3	Hopping	0.55
CH_2	Hopping	0.17
CH	Hopping	0.62
C	Hopping	0.72
H	Hopping	0.27

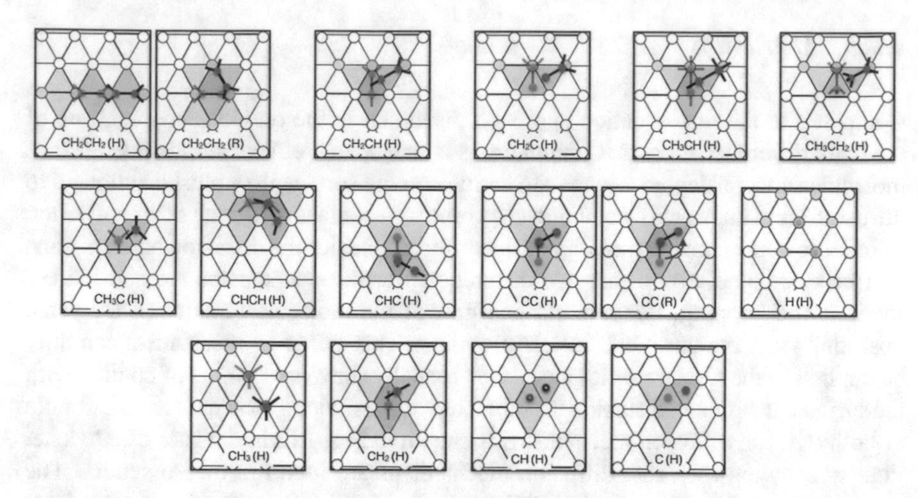

Fig. 4.14 The schematics for the diffusion of each of the species. The species diffuse by two different mechanisms: (H) hopping and (R) rotation. Each *box* shows one example of the diffusion process with the species moving from the *blue* to *green shaded* sites. The *light green shading* indicates the other possible sites the species can diffuse to. The sites where the product and reactant species overlap are indicated by the *dark green shading*

The diffusion barriers are all below 1 eV and therefore all the species are fairly mobile. In particular CHC and CH_2 have diffusion barriers less than 0.2 eV. H atoms also have a low barrier for diffusion (as previously noted in Sect. 3.4). The mobility of H atoms is important for hydrogenation reactions: since the reactions are bimolecular, they rely on H atoms being mobile enough to reach other species. This is also the case for the recombination reactions involving CH_3, CH_2, CH, and C. However since the diffusion barriers for these species are lower than the recombination reaction barriers, which all exceed 1 eV, their mobility should not limit these reactions. Therefore it is likely that none of the bimolecular reactions are diffusion limited.

Species diffusion can be treated within the kMC algorithm along with the other species. In this case the diffusion mechanisms for each species are considered with the other reactions, each with their own rate. The schematics for the diffusion of the species are shown in Fig. 4.14.

4.4.6 Product Species Fitting

For all reaction types (apart from H_2 desorption) the fitting of the product species on the surface must be checked in order to determine whether that reaction is allowed. To do this the reactant is temporality removed from the surface by setting the relevant grid sites to be unoccupied. Then, based on the calculated NEB reaction process, the positions the potential product species are generated. If any of the required sites for the new molecule are occupied then the product being checked cannot form in that position. If they are all unoccupied then both the reaction, and the possible species are added to the relevant lists.

4.5 Time Step Calculation

In each iteration of the kMC code the increase in the time must be calculated. This is the time that passes before a reaction move is made. For simulating the decomposition at a fixed temperature the rates are time independent and the time step can be calculated simply from the sum of the reaction rates R_{tot} (as explained in Sect. 2.7.3).

$$\Delta t = -\ln(RN) \times \frac{1}{R_{tot}} \tag{4.1}$$

where RN is a random number between 0 and 1. This ensures that the time step is chosen appropriately. When the temperature is ramped however, the rates become time dependent, since the temperature increases linearly with time and the rates are dependent on the temperature. In this case the time step is calculated from the integral of the total rates with respect to time.

$$\int_{t_{old}}^{t_{old}+\Delta t} R_{tot}(\tau)d\tau = -\ln(RN) \tag{4.2}$$

Note that the upper limit, Δt is unknown. The integral of R_{tot} is calculated from the current time in the simulation, t_{old} increasing the upper limit by a small amount, $d\tau$ until the integral is greater than $-\ln(RN)$. This criterion is fulfilled once the time step is large enough to ensure that a the probability of a reaction occurring is certain. Hence this time corresponds to Δt. Calculation of the time dependent time step in kMC is explained in detail in Sect. 2.7.3.1. The time dependent routine for calculating Δt is shown in Fig. 4.15.

Before determining the rate integral R_{int}, the small time interval for the integration $d\tau$ must be calculated. This needs to be sufficiently small compared to Δt. To ensure this the list of current reactions are checked to find the fastest move with the largest rate, $\max(R_i)$. Since this process would have the smallest time to occur $(= 1/\max(R_i))$, then taking $d\tau$ to be 1/100 of this value should give enough allowance to accurately calculate the integral.

In the next step the R_{int} is calculated self-consistently from $t_{old} \rightarrow t_{old} + t_j$ where $t_j = jd\tau$ and j is the number of integral steps. At each step the temperature T_j is calculated and then used to find the rates of each of the reactions. These are then summed and the integral is performed using Simpson's Method. If R_{int} is greater than $-\ln(RN)$ then $\Delta t = t_j$ and the integral is complete. Otherwise t_j is increased by $d\tau$ and the process repeats.

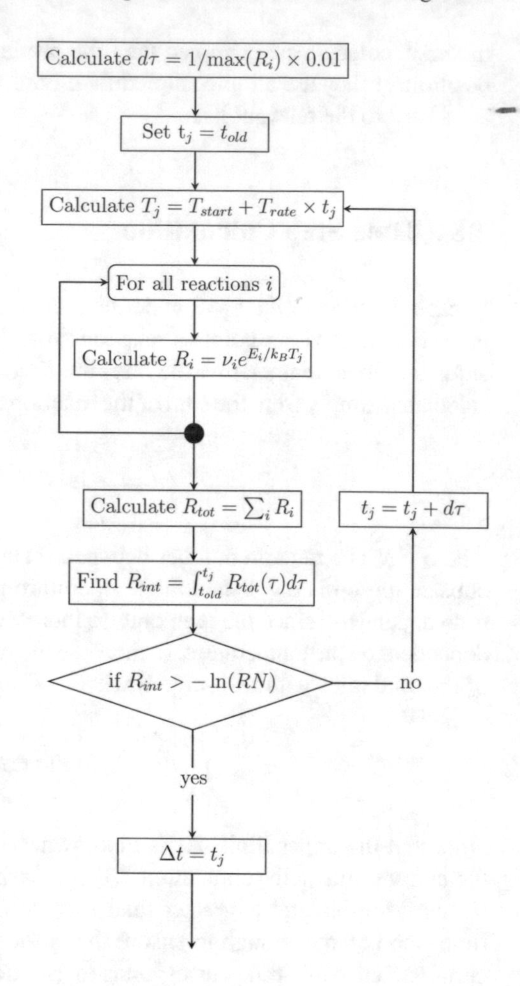

Fig. 4.15 The algorithm for calculating the variable time step when the rates are time dependent. Here T_{rate} is the rate of increase of the temperature

4.6 kMC Efficiency

The diffusion of most of the species can be considered as fast, high rate processes compared to the other reactions. This means that the system can easily get trapped if there are many species that have low diffusion barriers, simulating many diffusion moves each with a small time step. This increases the overall simulation time, and can become computationally impractical.

To avoid this, diffusion can be treated separately from the rest of the kMC moves. Note that this approach can only be implemented if the diffusion and the reaction processes can be split into two sets on the basis that the rates for diffusion are on average must faster than the rates for reaction processes. The simplest way to incorporate diffusion is to randomly move each species on the surface grid at the end of each reaction step in the kMC code. This assumes the species diffuse quickly relative to the reactions, such that the surface is completely homogenised after each reaction. To

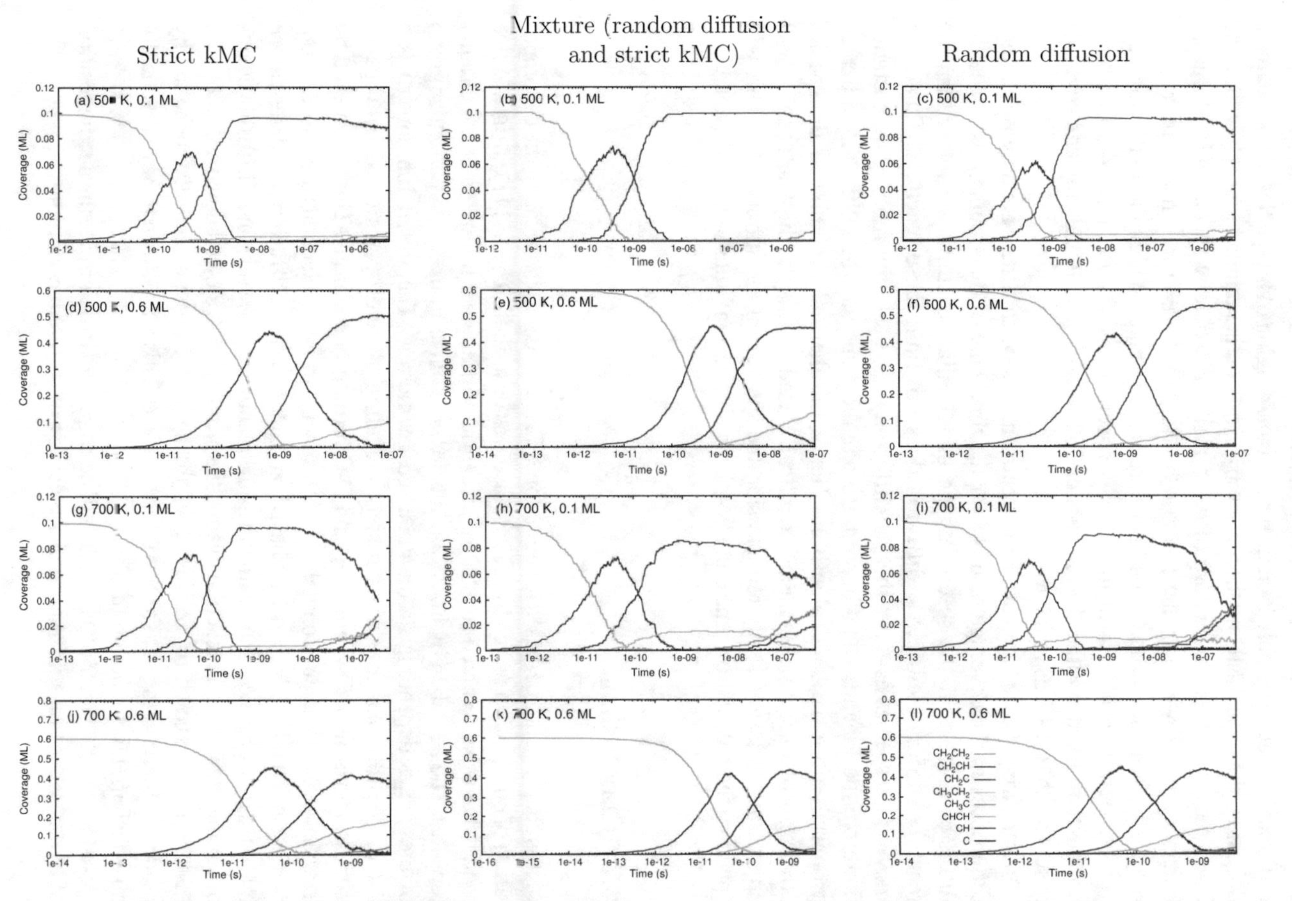

Fig. 4.16 The species evolution as determined by kMC simulations with strict kMC (*left figure*), a mixture of random and strict kMC (*centre figure*), and random diffusion (*right figure*) of species. The temperature and coverages are: **a**, **b**, **c** 500 K, 0.1 ML, **d**, **e**, **f** 500 K, 0.6 ML, **g**, **h**, **i** 700 K, 0.1 ML, **j**, **k**, **l** 700 K, 0.6 ML

check that this is a suitable approximation, the species coverages are calculated in a set of test simulations at constant temperature with diffusion incorporated using three different techniques. These include random rearranging of all species to approximate diffusion, the exact kMC method whereby diffusion is included properly within the kMC moves and also a mixture of these methods whereby species with diffusion barriers much lower than the reaction barriers (<0.4 eV) are treated by random reorganisation, and those with higher barriers are treated within the kMC moves.

The thermal evolution of the species coverages is shown in Fig. 4.16 with the temperature fixed at either 500 K (a-f) or 700 K (g-l). The initial ethylene coverage is set to either 0.1 ML or 0.6 ML.

Comparing the different diffusion treatments shows that there are only very minor differences in the coverage results over the timescale of the simulations. Since the diffusion barriers of the species of the bimolecular reactions are lower than the reaction barriers themselves, diffusion does not limit these reactions. Randomly moving the species assumes that the diffusion is fast and non-limiting to reactions, therefore the kinetics should be on average the same as when diffusion is treated properly within the kMC algorithm. Hence the thermal evolution of the species coverages in all regimes is almost identical. Based on this the diffusion can be well approximated by randomly reorganising the species on the surface grid. This will speed up the calculation time and allow long simulation times to be achieved.

4.7 Conclusions

In this chapter a lattice-based kinetic Monte Carlo model for simulating the decomposition of ethylene on an fcc Ir(111) surface was described. The Ir(111) surface lattice model that was presented consists of a 2D grid of triangular hollow sites and circular top sites. The adsorbed hydrocarbon species are coarse-grained to a particular group of these sites depending on their relaxed geometries. The simulation constructs a list of possible reactions at each step, including all hydrogenation, dehydrogenation, isomerisation, C-C breaking and C-C formation reactions. This is achieved by examining the local environment of each species in turn. Reactions are chosen from this list and the time is increased based on the total rates. Diffusion processes can be included by using the same treatment as the reactions; however, since diffusion events occur much faster than reaction events, the simulation is slowed down considerably. Instead it was shown that diffusion of the species can be effectively modelled by randomly rearranging the species on the surface. This was shown to have practically no effect on the simulation results.

In the next chapter the kMC code will be employed to determine the decomposition of hydrocarbons on substrates under a variety of different conditions.

Chapter 5
Thermal Decomposition in Graphene Growth: Kinetic Monte Carlo Results

As previously mentioned there are two main types of epitaxial graphene growth: temperature programmed growth (TPG) and chemical vapour deposition (CVD). The main difference between these is that in TPG the carbon source (such as hydrocarbon molecules) is deposited on a substrate at low temperatures before heating, whereas in CVD the deposition takes place once the system is already at a high temperature or during heating, by dosing the substrate with hydrocarbons. Usually this happens continuously throughout the growth. In this section the decomposition of ethylene on the Ir(111) surface will be simulated with the kMC code as if graphene were to be produced by both of these techniques. Within temperature programmed growth both the decomposition mechanism due to ramping of the temperature and the decomposition mechanism at a fixed temperature will be determined.

In addition to the experimental method, the ethylene decomposition process may also be affected by the choice of growth substrate. The substrate interaction with the adsorbed hydrocarbon species will affect the energy barriers for the reactions. As well as this there may be configurational differences in the relaxed geometries of the species, and the mechanisms for both the reactions and the diffusion of species. To demonstrate this the decomposition mechanism of ethylene on the Pt(111) surface, as well as the Ir(111) surface, will be simulated. This will be modelled based on the hydrocarbon geometries and the reaction energy barriers reported in [1]. By comparing the mechanism with that of ethylene decomposition on Ir(111), differences between the early stages of graphene growth on these two substrates may be uncovered.

Finally, the choice of growth precursor itself will have an effect on the temperature required to form C monomers (or other C_n species), and hence affect the temperature for graphene growth. This was demonstrated in [2]. The thermal decomposition of methane (CH_4) and ethylene (CH_2CH_2) in CVD growth will be compared using kMC simulations.

H.A. Tetlow, *Theoretical Modeling of Epitaxial Graphene Growth on the Ir(111) Surface*, Springer Theses, DOI 10.1007/978-3-319-65972-5_5

5.1 Temperature Ramping Programmed Growth

5.1.1 kMC Results

The decomposition of ethylene in the early stages of the temperature ramping programmed growth of graphene (as experimentally determined in Chap. 3) is simulated. The ethylene molecules are deposited onto the Ir(111) surface at low temperatures and then the system is heated gradually, which causes ethylene to decompose.

The thermal evolution of ethylene decomposition as determined by the kMC simulations is shown in Fig. 5.1. The temperature is ramped from 100 K by 1.5 Ks^{-1}. Two different initial ethylene surface coverages are shown: (a) 0.05 ML and (b) 0.6 ML.

Each of the simulations with varying coverage have a similar sequence of species appearing. Initially ethylene dehydrogenates to CH_2C, and, eventually after various

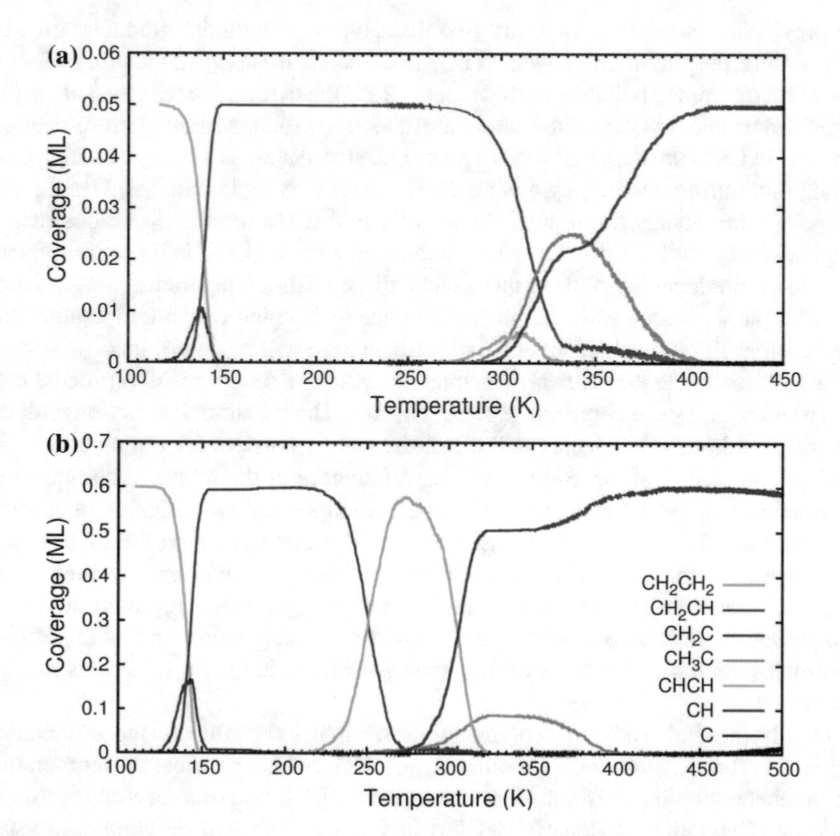

Fig. 5.1 The thermal evolution of ethylene decomposition on the Ir(111) surface with a temperature ramp of 1.5 Ks^{-1}. The initial coverage of ethylene is **a** 0.05 ML and **b** 0.6 ML. Adapted with permission from [3]

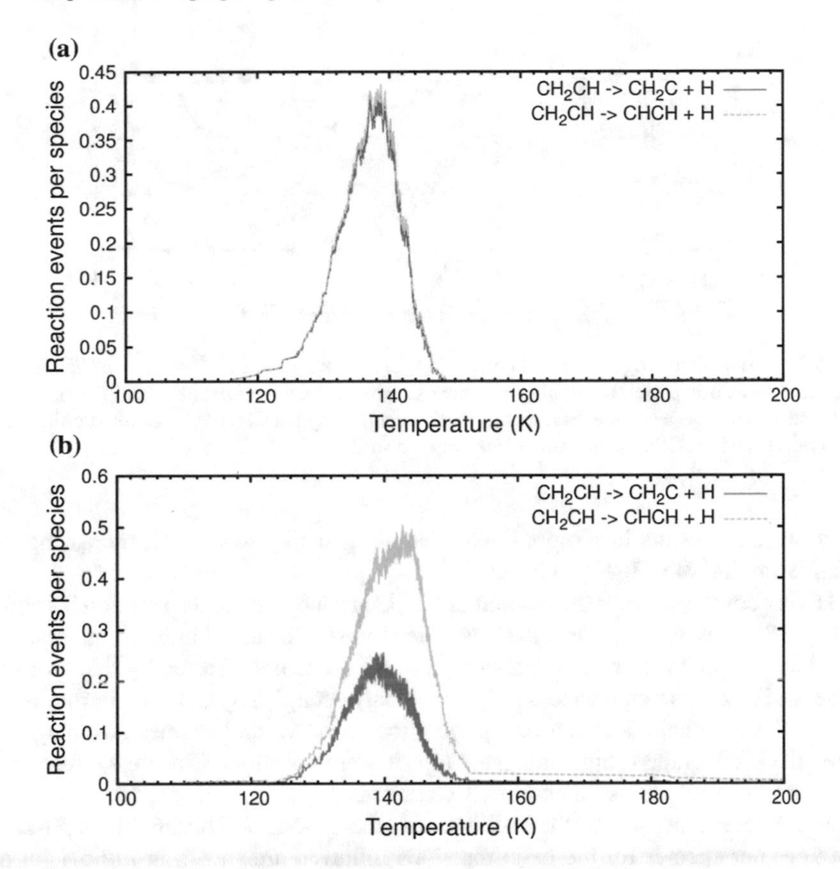

Fig. 5.2 The number of possible reactions found for the CH_2CH dehydrogenation reactions during the kMC simulations for the two coverages shown in Fig. 5.1 **a** 0.05 ML and **b** 0.6 ML

intermediate species are formed, the C-C bond must be is broken to give CH and then C monomers. This is in agreement with the experimental results (Fig. 3.7) and the rate equations (Fig. 3.11) in Chap. 3.

Based on these results the effect of increasing coverages can be seen on the thermal evolution. At the coverage of 0.05 ML CH_3C is the most prominent intermediate species over the temperature range of 300–400 K with CHCH showing only a small peak at 300 K. As the coverage is increased to 0.6 ML the amount of CHCH grows, and the amount of CH_3C decreases. Correspondingly, the disappearance of CH_2C is shifted to lower temperatures since it converts to CHCH at lower temperatures than CH_3C.

These coverage effects can be explained by the number of possible reactions per species for the CH_2CH dehydrogenation reactions as shown in Fig. 5.2. As the coverage is increased the number of possible reactions for CH_2CH dehydrogenation to CH_2C decreases. This means that, according to the simulations based on DFT calculated rates, that increasing the surface coverage limits the $CH_2CH \rightarrow CH_2C + H$

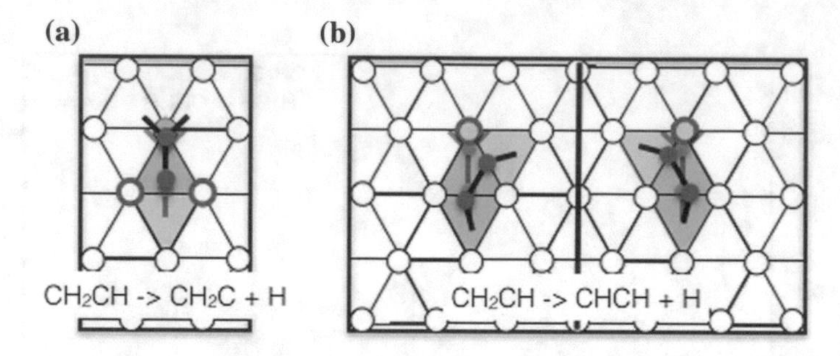

Fig. 5.3 The two dehydrogenation reactions of CH_2CH resulting in **a** $CH_2C + H$ and **b** $CHCH + H$. The reaction in **b** is preferable in high coverage situations since the *top* site where the H atom is removed to is always available, as it was formerly occupied by the CH_2CH molecule. Furthermore the product species, CHCH does not occupy any *top* sites

reaction. This results in more CH_2CH dehydrogenating to CHCH, increasing the relative amount of CHCH produced.

Higher coverages limit the amount of CH_2C produced since the reaction 3, shown in Fig. 5.3a, is only possible if there is a free top site for the H atom to be removed to. The reaction 14 (in reverse), shown in Fig. 5.3b, is not affected by this problem since the H atom is removed to the site previously occupied by CH_2CH. Furthermore the CHCH molecule does not occupy any top sites, which becomes advantageous when the H coverage is high and many top sites are occupied. Coverage effects will therefore increase the likelihood of CHCH forming.

The coverage also affects the likelihood of other reactions. The number of possible reactions per species for the main reactions followed are shown in Fig. 5.4 for an initial coverage of (a) 0.05 ML and (b) 0.6 ML.

The high coverage results for the number of possible reactions can be seen to follow the species coverage results in Fig. 5.1b. When a species is produced in high quantities the reactions which it can undergo become possible. This is seen across all temperature ranges. For the low coverage results the relative number of the possible reactions are quite different. There is a large reduction in the number of possible hydrogenation reactions (and bimolecular reactions) due to the requirement for the two species to be located near to each other for the reaction to be allowed. Furthermore the number of dehydrogenation reactions are increased since there are many free sites for the H atoms to occupy after the reaction.

5.1.2 *Comparison with Experimental Results*

The experimental results at high and low initial ethylene coverages are shown in Fig. 5.5a and b respectively.

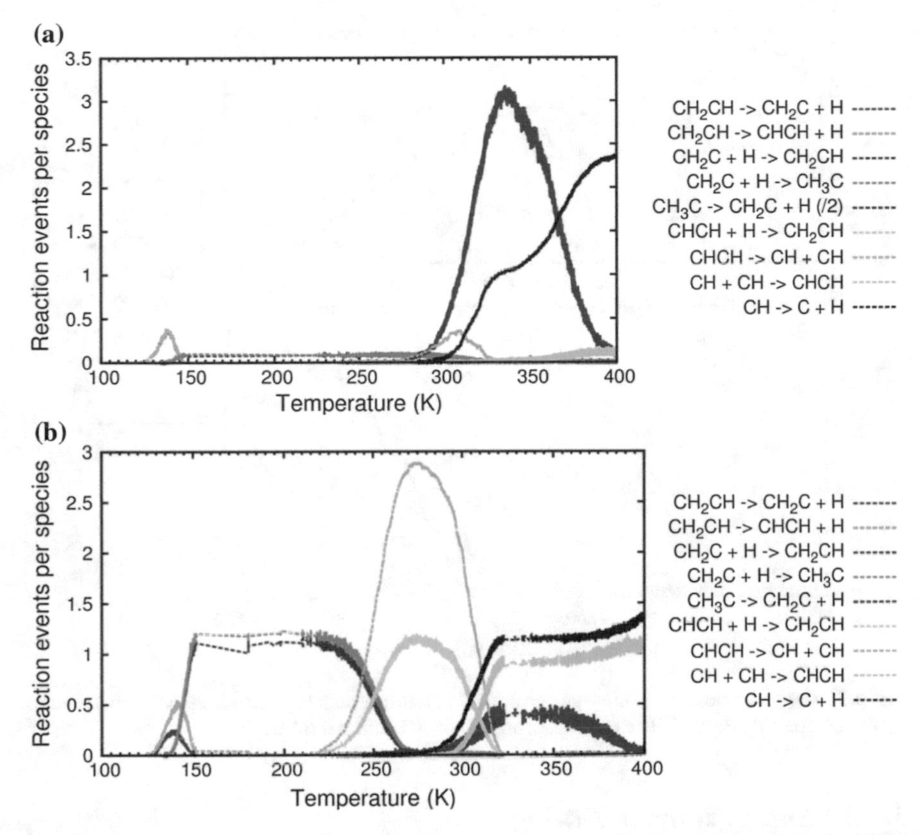

Fig. 5.4 The number of possible reactions found for the main reactions during the kMC simulations for the two coverages shown in Fig. 5.1: **a** 0.05 ML and **b** 0.6 ML

At high coverages the kMC deduced thermal evolution of species agrees qualitatively with the experimental results. However there are noticeable differences in the temperature ranges when certain species appear, as demonstrated by the temperature windows for the various species in the kMC results being narrower compared to the experimental results.

At low coverages the agreement with the experimental results is weaker since the kMC simulations predict a greater amount of CH_3C and less CHCH over the temperature range of 250–450 K. Overall however the effect of increasing coverage is similar for both experimental and theoretical results: CHCH is more prevalent at higher coverages whereas at lower coverages more CH_3C is formed.

Since the kMC simulations do not match up with the experimental results, especially at low coverages when interactions between species should be minimal and hence the calculated rates should be more reliable, it is possible that the reaction rates are affected by systematic errors in the DFT calculated energy barriers.

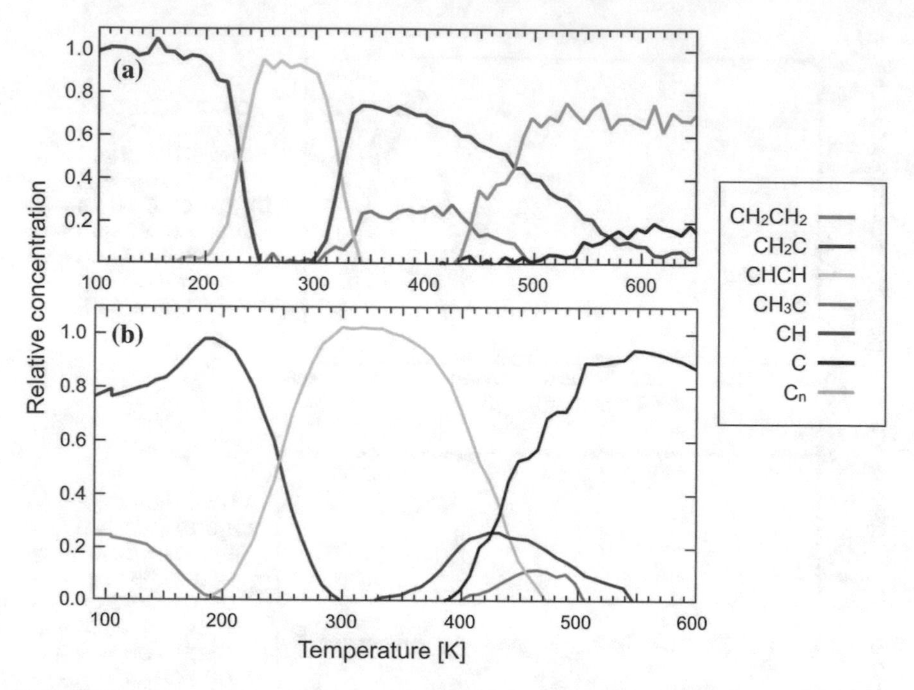

Fig. 5.5 The decomposition of ethylene on the Ir(111) surface as determined from XPS experiments by the group of A. Baraldi. The coverages are **a** 0.05 ML and **b** 0.6 ML

5.1.3 Energy Barrier Tuning

Due to the inherent error in the DFT calculated energy barriers the actual value of the barriers may vary by 5–10% from their calculated values [4]. This may cause fairly significant errors in the species evolution as determined from the kMC simulations. This will be particularly problematic when different competing reaction mechanisms have similar energy barriers, which is the case for this particular reaction system. Once CH_2C has been produced from ethylene dehydrogenation, the reaction processes converting it to either CHCH or CH_3C have energy barriers which are almost identical. As shown in the energy barrier profile in Fig. 5.6, hydrogenation to CH_3C has a barrier of 0.82 eV, whereas the total energy increase for converting CH_2C to CHCH (via (d) → (c) → (b)) is 0.84 eV (assuming CH_2CH is very short lived). Therefore small differences in these barriers may change the preferred reaction mechanisms drastically.

By adjusting these energy barriers by small amounts it is possible to get a much better agreement with the experimental results. To achieve this the energy barrier for (d) → (e) in Fig. 5.6 can be increased slightly, and the barriers for (d) → (c) and (c) → (b) can be decreased, hence suppressing the formation of CH_3C, and promoting the formation of CHCH. The barrier for reaction from (b) → (a) can also be increased from 0.85 to 0.90 eV in order to postpone the formation of CH and hence increase

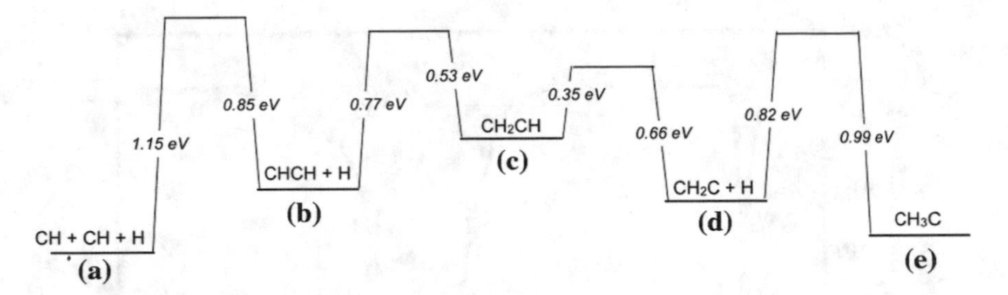

Fig. 5.6 The energy profile of the various important competing reaction processes from CH_2CH to either, CH_3C (*right*) or CHCH (*left*), illustrating the differences in barrier heights

Table 5.1 The adjustments made to the energy barriers from the calculated values

	Reaction	E_{calc} [eV]	E_{adjust} [eV]
3_{rev}	$CH_2C + H \rightarrow CH_2CH$	0.66	0.58
7_{for}	$CH_2C + H \rightarrow CH_3C$	0.82	0.85
14_{rev}	$CH_2CH \rightarrow CHCH + H$	0.53	0.48
$CB6_{for}$	$CHCH \rightarrow CH + CH$	0.85	0.90

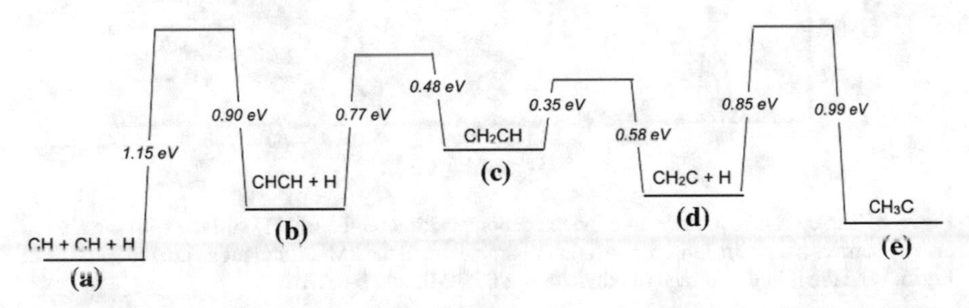

Fig. 5.7 The energy profile of the reaction processes with the energy barriers altered as indicated in Table 5.1. The barriers have been altered so that the barrier from **d** to **e** is comparatively higher than the "effective" barrier from **d** to **b**

the temperature range for which CHCH survives. The changes made to the energy barriers as compared to the calculated values in Table 3.2 are shown in Table 5.1.

By making these (rather small) adjustments a new energy profile is constructed. This is shown in Fig. 5.7.

It is also of interest that at high coverages the experimental results show a general broadening to the temperature ranges of the CH_2C and CHCH species. This suggests that at higher coverages the lifetimes of the species are relatively increased. This may be due to the increased effect of interactions between species at higher coverages, which can affect the energy barriers. These effects cannot be modelled within the kMC simulation since it would require the inclusion of an exhaustive number of different reaction scenarios corresponding to reactions in various environments. Instead interactions may be incorporated by scaling up the values of energy barriers for high coverage simulations.

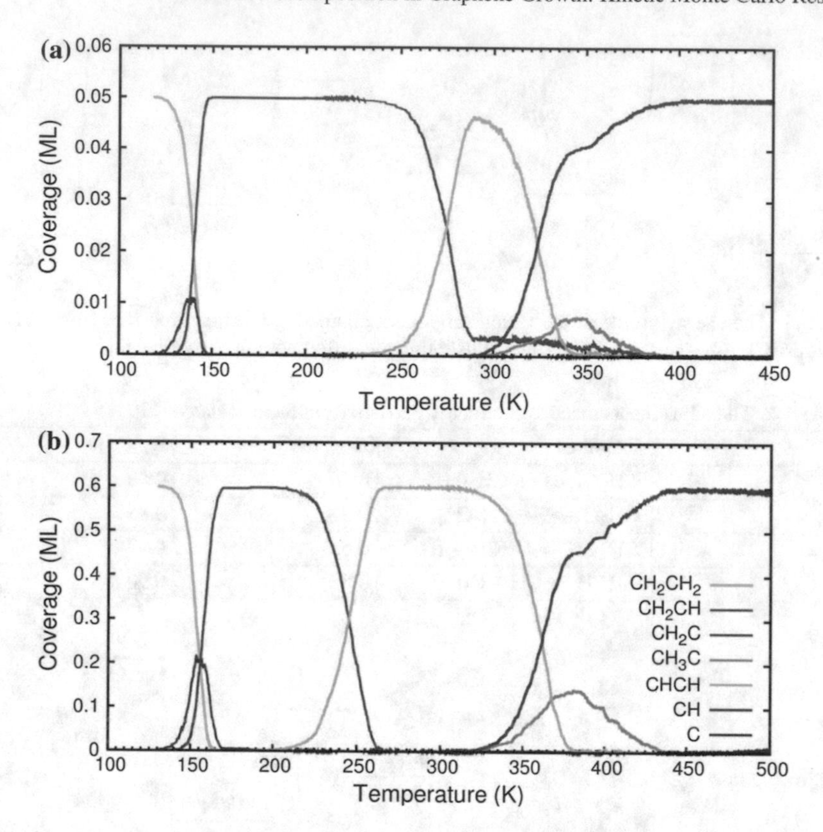

Fig. 5.8 The thermal evolution of ethylene decomposition on the Ir(111) surface with the altered energy barriers shown in Table 5.1. All other parameters in the kMC simulations are the same as in Fig. 5.1 and the initial coverage of ethylene is **a** 0.05 ML and **b** 0.6 ML

Figure 5.8a shows the low coverage kMC results with the energy barrier adjustments in Table 5.1 applied. CHCH is now the main species between 250 and 350 K. The results agree with the experimental species at low coverages in Fig. 5.5a, apart for the initial temperature range where CH_2C and CHCH form, which is higher in both cases. The high coverage kMC results, with a +10% scaling to the (adjusted) energy barriers applied, contain the same sequence of species as in the experiment and the results without any energy barrier adjustment. The main difference as a result of the adjustments is that the temperature ranges where each species is present are comparatively larger. This is a better comparison with the experimental results, which predict a large temperature window for CHCH, from 200–450 K.

Overall the experimentally determined species evolution and the kMC simulated species evolution agree qualitatively with each other. In both cases the same reaction pathway is obtained despite the fact that the ratios of intermediate species may be different. Given that the exact experimental conditions, including specific species interactions, cannot be explicitly included into the kMC model, an exact quantitative agreement may be difficult to achieve. Furthermore the DFT calculations for the energy barriers cannot predict the values exactly, which due to the delicate nature of the specific reaction processes studied, means that the relative coverages of some of the species may vary slightly.

5.1.4 Comparison with Rate Equations

The decomposition kinetics can also be modelled by rate equations, as demonstrated previously in Chap. 3. Constructing rate equations is far simpler than a kMC code, since all that is required are the reactions and their rates. Solving rate equations is also much faster computationally as it involves only solving a set of simultaneous differential equations. A disadvantage is that rate equations contain no spatial information about the processes, and coverage effects such as diffusion, are not included. However, despite this, there are some cases where rate equations can still meaningfully approximate the species evolution.

The kMC results for ethylene decomposition can be compared directly to the species coverages determined by the time evolution of the rate equations. Two different initial ethylene coverages, 0.05 ML and 0.6 ML, are considered. The evolution of hydrocarbon species as calculated by rate equations is shown in Fig. 5.9. The energy barriers correspond to the adjusted values in Table 5.1.

At the lowest coverage, 0.05 ML, the rate equation results and the kMC results are in good agreement. Initially ethylene dehydrogenates to CH_2C at 150 K. At around 250 K the CH_2C is converted into CHCH. At 300 K CH begins to form, along with a small amount of CH_3C. This survives until 400 K. The CH molecules eventually dehydrogenate to C monomers above 500 K (not shown in the kMC results).

At a coverage of 0.6 ML the kMC and rate equation results differ in several ways. The rate equation results at high coverage resemble those at the lower coverage, whereas the high coverage kMC results predict a much longer lifetime for CHCH, forming at 200 K and existing until 380 K, and correspondingly a much shorter lifetime for CH_2C. This is due to the increasing effect of the coverage on the kMC reactions. As mentioned previously, increasing the coverage has a positive effect on the amount of CHCH formed. At high coverages, when the surface is saturated with H atoms, dehydrogenation to CHCH becomes more preferable than dehydration to CH_2C, due to the fact that reaction 3, shown in Fig. 5.3a, is only possible if there is a free top site for the H atom to be removed to.

At low coverages (0.05 ML) both the rate equations and kMC simulations give results which show a good agreement with the experimental results in Fig. 5.5a. In addition, the kMC results at high coverages (0.6 ML) with adjusted energy barriers show a close agreement to the experimental results in Fig. 5.5b. Both indicate CHCH is the predominate species between 280 and 350 K. However the rate equations fail at high surface coverages since they cannot account for any spatial effects. Therefore rate equations may provide a good approximation to the ethylene thermal evolution of ethylene at low coverages.

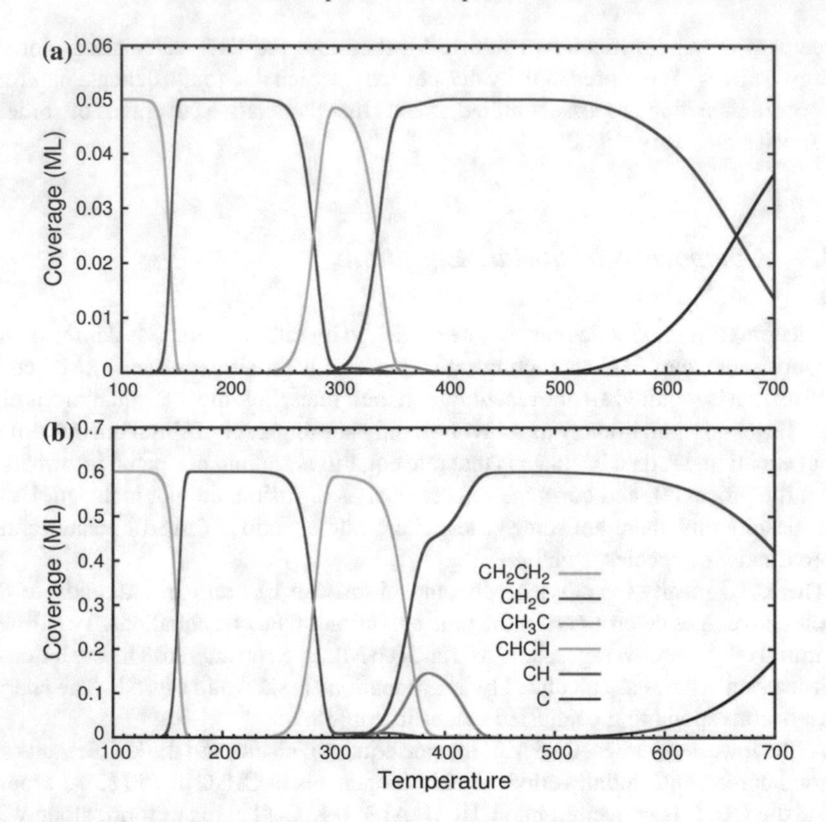

Fig. 5.9 The thermal evolution of ethylene decomposition on the Ir(111) surface as determined from the solution of rate equations. The initial coverage of ethylene is **a** 0.05 ML and **b** 0.6 ML

5.2 Fixed Temperature Programmed Growth (kMC)

In addition to temperature ramping programmed growth discussed above, the ethylene decomposition process, and the graphene formation that follows, can also be facilitated by heating the substrate to a fixed arbitrary temperature for a fixed amount of time. This was shown for graphene growth on the Ir(111) surface using ethylene as a precursor [5]. Higher temperatures produced more advanced, and larger graphene islands.

The decomposition at the fixed temperatures of 300, 400, 500, and 700 K as calculated with kMC is shown in Fig. 5.10a–h. For the low temperature results the first 60 seconds of the simulation is shown since heating periods do not exceed this in graphene growth experiments.

For both high and low coverages and for all temperatures the evolution of the species appearing on the surface is similar to when the temperature is linearly ramped, as in Sect. 5.1. The reaction sequence begins with dehydrogenation of ethylene to

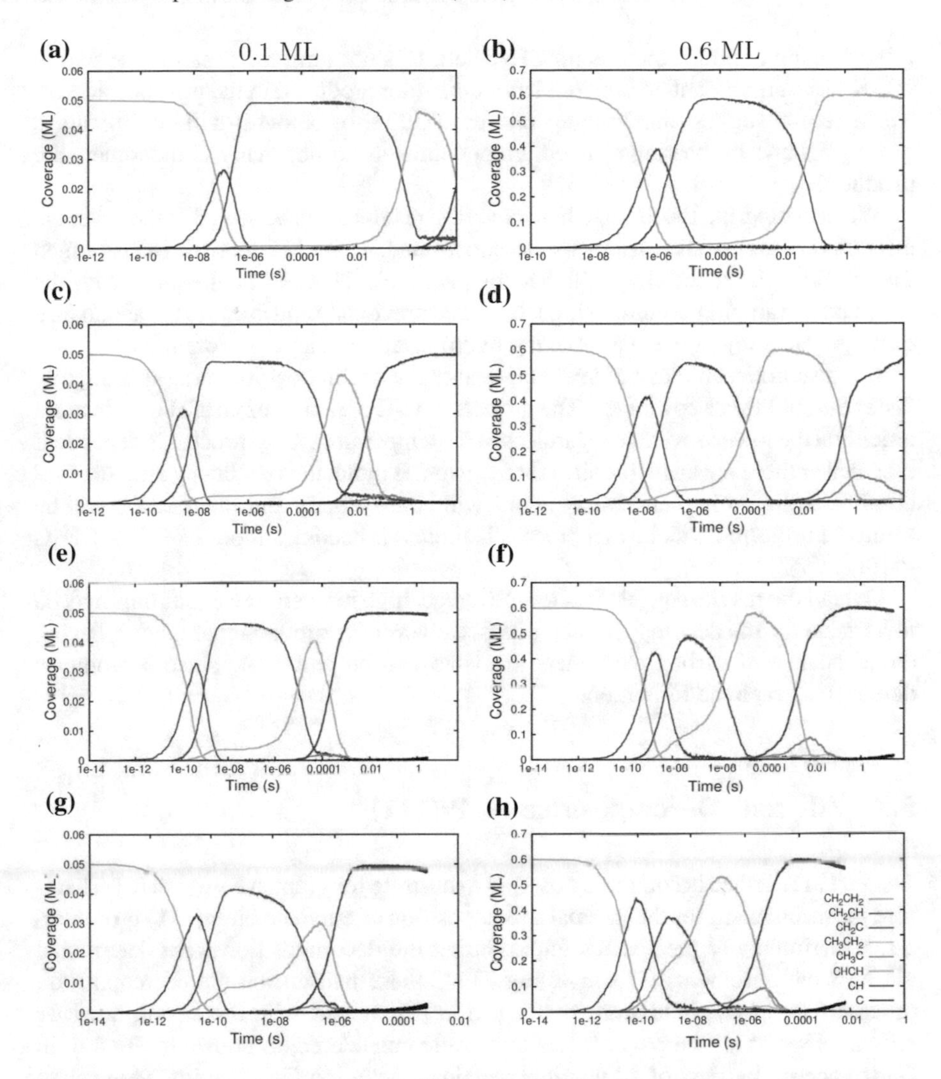

Fig. 5.10 The ethylene decomposition at different fixed temperatures and coverages, as calculated using the kMC code. **a** 300 K, 0.1 ML, **b** 300 K, 0.6 ML, **c** 400 K, 0.1 ML, **d** 400 K, 0.6 ML, **e** 500 K, 0.1 ML, **f** 500 K, 0.6 ML, **g** 700 K, 0.1 ML, **h** 700 K, 0.6 ML

CH_2CH and then CH_2C. This converts to CHCH, which then divides into two CH molecules, before dehydrogenating to C monomers. Some CH_3C may be formed depending on the conditions.

As the heating temperature is raised the thermal evolution of the species speeds up and at the end of the heating period different species are present. The simulation at 300 K ends with CHCH being the main species on the surface. At 400 K the evolution has progressed further and CH is the main species, with some CH_3C also present.

After heating at 500 K there is no CH_3C left. It is not until the system is heated to 700 K that a significant amount of C monomers are produced in the time considered. These results suggest that at temperatures of 500 K or below it will be difficult to grow graphene by heating at fixed temperatures since not many C monomers are produced.

When comparing the low and high coverage results it can be seen that the coverage affects how much of the various species are formed. At 0.6 ML more CH_3C is formed than at 0.1 ML, especially at higher temperatures. This is in agreement with the temperature ramping decomposition discussed previously and is due to the abundance of hydrogen at high coverages making hydrogenation reactions more likely.

It is also noticeable that at lower coverages the decomposition process is slightly faster than at higher coverages. The process of C-C breaking of CHCH occurs more quickly at the lower coverage regardless of the temperature. This process is dependent on whether there are available sites for the two CH molecules to occupy after the C-C bond has split. At higher coverages this will be less likely and the reaction will be limited. Furthermore since this process is limited it becomes more likely for CH_3C to form.

Overall the results suggest that low coverage, high temperature conditions may be favourable for the decomposition process. However for producing graphene, having an abundance of carbon monomers available will be preferable since it promotes defect-free graphene formation.

5.3 Ethylene Decomposition on Pt(111)

The Pt(111) surface is commonly used as a substrate for graphene growth [6]. Therefore the mechanism for the thermal decomposition of ethylene on Pt(111) is of interest. Experimentally the species found during the decomposition were determined via XPS measurements [7] suggesting CH_3C is formed during the decomposition process. The kMC simulations performed in [8], agreed with this result, however only a subset of all the possible reactions were considered, as shown in Fig. 1.6. In [1] the energy barriers of all possible reactions including C-C scission were calculated, however the reaction kinetics was not determined.

In this section the kinetics for ethylene decomposition on Pt(111) based on the reaction energy barriers in [1] will be simulated. Since Pt(111) and Ir(111) are similar surfaces, with the hydrocarbon species adopting the same geometries on both, only the energy barriers (and pre-exponential factors) need to be replaced in the kMC code input in order to simulate ethylene decomposition on Pt(111). Instead of diffusing each species within the kMC method the species are randomly rearranged at each time step as in Sect. 4.6.

5.3.1 Energy Barriers

The energy barriers for the reactions in Fig. 3.1 on both the Ir(111) and Pt(111) surfaces are shown in Table 5.2. The Pt(111) barriers are taken from [1] where the calculations were performed using a constrained optimisation scheme with the PBE functional. Compared to the energy barriers for the equivalent reactions on Ir(111), the Pt(111) barriers are on average larger. An exception to this is reaction 8 where CH_3CH dehydrogenates to CH_3C on Pt(111) with a 0.28 eV barrier, compared to 0.46

Table 5.2 The energy barriers for the various reactions numbered as in Fig. 3.1. E_f is the energy barrier associated with the forward direction of the arrow, whereas E_b is the backward reaction barrier. The Pt(111) barriers are taken from [1]

	Reaction in forward direction	E_f^{Ir} [eV]	E_b^{Ir} [eV]	E_f^{Pt} [eV]	E_b^{Pt} [eV]
1	$CH_2CH_2 \rightarrow CH_2CH + H$	0.39	0.58	0.84	0.82
2	$CH_2CH_2 + H \rightarrow CH_3CH_2$	0.70	0.35	1.12	0.81
3	$CH_2CH \rightarrow CH_2C + H$	0.35	0.66	0.70	0.90
4	$CH_2CH + H \rightarrow CH_3CH$	0.64	0.27	1.02	0.71
5	$CH_3CH_2 \rightarrow CH_3CH + H$	0.33	0.49	0.88	0.87
6	$CH_3CH_2 + H \rightarrow CH_3CH_3$	2.16	–	1.02	–
7	$CH_2C + H \rightarrow CH_3C$	0.84	0.99	0.83	1.33
8	$CH_3CH \rightarrow CH_3C + H$	0.46	1.48	0.28	1.13
9	$CH_2CH_2 \rightarrow CH_3CH$	1.39	1.20	2.36	2.04
10	$CH_2CH \rightarrow CH_3C$	1.37	2.01	1.96	2.50
11	$CHC + H \rightarrow CH_2C$	0.54	1.17	1.10	2.22
12	$CHC \rightarrow C\text{-}C + H$	1.23	0.65	–	–
13	$CHCH \rightarrow CHC + H$	1.23	0.58	2.12	1.22
14	$CHCH + H \rightarrow CH_2CH$	0.77	0.53	1.01	1.03
15	$CHCH \rightarrow CH_2C$	2.44	2.52	2.60	2.38
16	$CH_3 \rightarrow CH_2 + H$	0.50	0.58	0.83	0.76
17	$CH_2 \rightarrow CH + H$	0.09	0.83	0.17	0.82
18	$CH \rightarrow C + H$	1.11	0.66	1.29	0.82
CB1	$CH_2CH_2 \rightarrow CH_2 + CH_2$	1.45	0.61	2.22	1.59
CB2	$CH_3CH_2 \rightarrow CH_3 + CH_2$	1.56	1.47	1.84	1.59
CB3	$CH_3CH \rightarrow CH_3 + CH$	0.89	1.31	1.18	1.59
CB4	$CH_2CH \rightarrow CH_2 + CH$	1.07	1.44	1.70	1.74
CB5	$CH_2C \rightarrow CH_2 + C$	1.88	0.80	2.22	0.8
CB6	$CHCH \rightarrow CH + CH$	0.79–0.93	1.15	1.07	1.78
CB7	$CHC \rightarrow CH + C$	0.73	1.29	0.90	2.04
CB8	$CC \rightarrow C + C$	1.18	1.29	–	–
H	$H + H \rightarrow H_2$	1.25	–	0.91	0.58

eV on Ir(111). The preferred direction of the reactions, which depends on whether E_f or E_b is larger, is the same on both substrates. This suggests that the decomposition mechanism should be expected to be similar to that on Ir(111). However the equivalent decomposition reaction processes may occur at higher temperatures than on Ir(111) as the Pt(111) energy barriers are larger.

5.3.2 kMC Results

The ethylene decomposition on Pt(111) is shown in Fig. 5.11a. At 250 K ethylene begins to dehydrogenate to CH_2CH and then CH_2C, producing hydrogen, as shown in Fig. 5.11b. Some of this hydrogen is used up in the hydrogenation of CH_2C to CH_3C, however most of it is removed from the surface via the formation of H_2 gas. The system remains with about 90% CH_2C and 10% CH_3C until around 450 K as CH_3C cannot form easily once most of the H has been lost from the surface. At

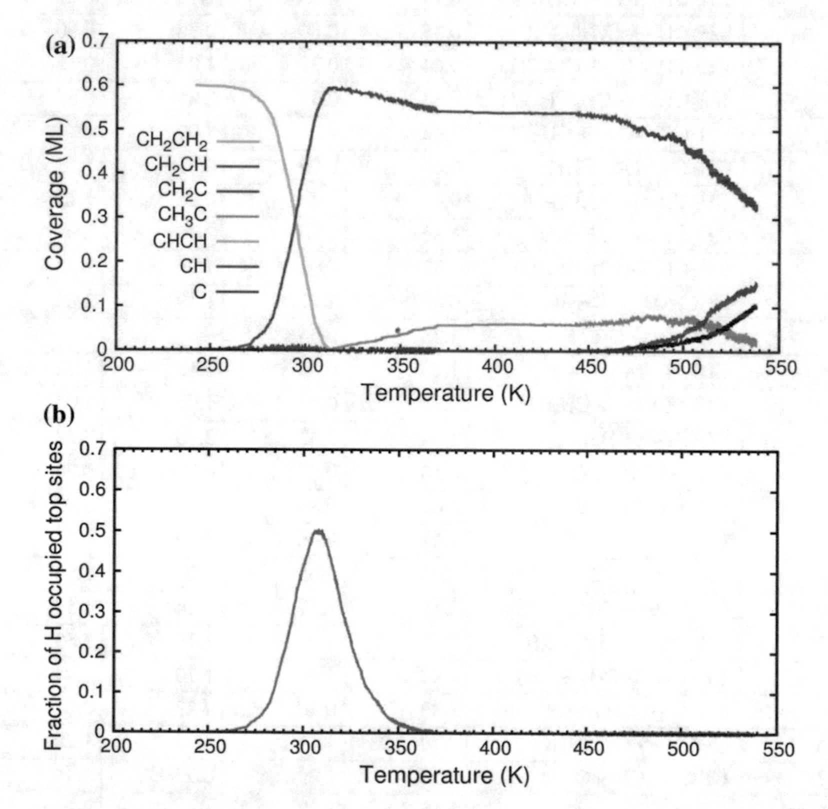

Fig. 5.11 The thermal decomposition of ethylene on Pt(111) showing **a** the hydrocarbon species on the surface and **b** the relative amount of H atoms. The energy barriers used in the kMC simulation are taken from [1]

about 450 K CH and C begin to form, accompanied by a reduction in the amount of both CH_2C and CH_3C. This occurs as CH_2C slowly converts to CHCH (via the path: $CH_2C + H \rightarrow CH_2CH \rightarrow CHCH + H$). The energy barrier for CHCH to split into two CH molecules is low compared to the other C-C breaking reactions (1.07 eV), so this process is followed above 450 K. Below this temperature the CHCH would convert back to CH_2C. Once formed, some of the CH dehydrogenates to C monomers. Unlike on the Ir(111) surface this process does not reverse easily since there is a limited amount of H on the surface at these temperatures.

The decomposition pathway for ethylene to C monomers on the Pt(111) surface is exactly the same as on the Ir(111) surface. Differences occur in which processes are limiting to the decomposition to C monomers. For the Ir(111) surface this is the removal of H atoms from the surface, which causes CH to rehydrogenate after dehydrogenating to C monomers, and prevents C monomers forming in large numbers. In the case of the Pt(111) surface, breaking of the C-C bond in CHCH has a higher energy barrier, which means that CH and C can only form above 450 K. There is also a limited amount of H on the surface above 350 K, and therefore the conversion of CH_2C to CHCH will be restricted as this requires CH_2C to hydrogenate to CH_2CH first. As a result, CH_2C and CH_3C coexist over a large temperature interval, before CHCH stars to form, which then immediately decomposes.

5.4 Chemical Vapour Deposition

Chemical vapour decomposition differs from temperature programmed growth, in which the hydrocarbon molecules are assumed to be already present on the surface prior to the temperature ramp. In CVD the hydrocarbon molecules are deposited onto the substrate all the way throughout the heating process. The temperature, like in TPG growth, may be kept at a fixed constant value or ramped continuously. In this section the thermal decomposition of ethylene and methane when deposited as in CVD (with a temperature ramp) will be simulated using the kMC code.

In order to simulate CVD within the kMC code a reaction to represent the deposition of species must be added. The rate from this process will differ from the other reactions since its pre-exponential factor will depend on the probability for the molecule to adsorb onto the surface. The rate for a single species i to adsorb onto a free surface site is given by,

$$R_{ads,i} = P_i A_{site,i} (2\pi M_i k_B T_{gas})^{-0.5} \exp(-E_i / k_B T) \tag{5.1}$$

where P_i is the partial pressure of the species i, A_{site} is the area of an adsorption site for species i, M_i is the mass of species i in kg [9], and T_{gas} is the temperature of the gas (room temperature). E_i is the energy barrier for the adsorption and T is the temperature of the metal surface. The adsorption reactions are implemented into the kMC code as a separate reaction type, whereby each empty surface site contributes one reaction event with a rate $R_{ads,i}$.

The kMC simulation begins with a bare Ir surface grid. At each reaction step the surface sites are scanned and for each unoccupied species adsorption site it is checked whether the species can adsorb, given its local environment. If this is possible then an adsorption reaction is added to the reaction list with a rate that is dependent on the temperature. The choice of parameters is discussed below. For ethylene all reactions in Fig. 3.1 are included. For the decomposition of methane the reactions included are the hydrogenation and dehydrogenation of the CH_m species, including the hydrogenation of CH_3 to methane gas.

5.4.1 Ethylene

For ethylene adsorption the partial pressure $P_i = 5 \times 10^{-10}$ mbar in Eq. (5.1), in accordance with the experimental conditions for graphene growth on Ir(111) [5]. The area of the absorption site is equal to twice the area of a hollow site of the fcc lattice $A_{site} = \sqrt{3}D$, where D is the metal-metal atomic distance. This area is based on the sites ethylene occupies as shown in Fig. 4.3. T_{gas} is the gas temperature and is equal to room temperature (300 K). The energy barrier for ethylene adsorption on Ir(111) is negligible. Based on these values the rate of ethylene absorption per available site is $1.56 \times 10^{-3} T_{gas}^{-0.5}$ s^{-1}.

The thermal evolution of the species during constant ethylene CVD deposition as the temperature is ramped at 1.5 Ks^{-1} is shown in Fig. 5.12. The energy barriers are adjusted as in Table 5.1.

Figure 5.12 shows that overall ethylene decomposition mechanism during CVD growth does not vary from the mechanism in TPG growth. When compared to the ethylene TPG decomposition, shown in Fig. 5.8a, the main difference is the overall increase in the total concentration of species as the temperature is raised. This reaches a total coverage of 0.12 ML. There is also noticeably less CHCH and more CH_3C formed than in the TPG results in Fig. 5.8. This is produced from additional ethylene that adsorbs above 300 K, as the CHCH begins to convert into CH at 300 K. The additional ethylene immediately dehydrogenates to CH_2C and then hydrogenates to CH_3C as there are many hydrogen atoms and also enough thermal energy to overcome the 0.84 eV barrier. CH_3C remains until above 400 K when it can dehydrogenate to CH_2C + H, and then CH is formed via the usual route with CHCH being an intermediate.

Another difference resulting from CVD with ethylene is that C monomers are formed at a lower temperature (450 K compared with 500 K for TPG growth). Unlike in TPG decomposition the C monomers begin forming before the majority of the H atoms have been lost to the gas phase, see Fig. 5.12b. This may be due to the constant flux of ethylene molecules which adds additional species to the surface. Above 400 K the ethylene will immediately decompose into two CH molecules and two H atoms. As some of the H atoms will desorb from the surface, this increases the relative amount of CH on the surface, and hence there is a greater concentration

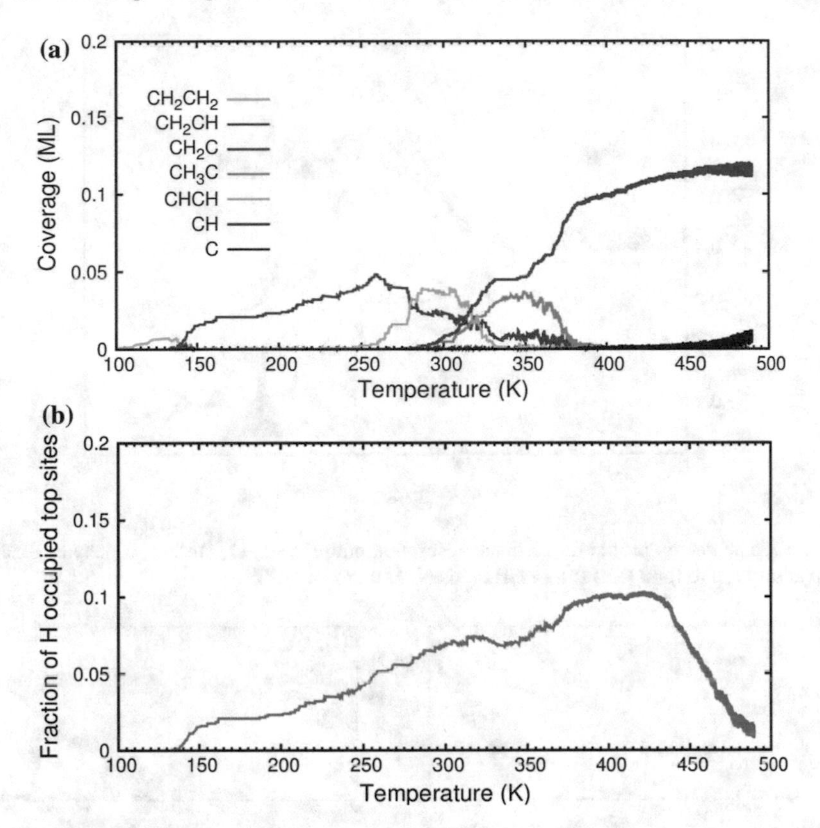

Fig. 5.12 **a** The thermal evolution of species on the Ir(111) surface, while ethylene is constantly dosed into the system and **b** the relative amount of H atoms

of C monomers being formed. This may be preferable to the decomposition during TPG as the C monomers are formed at a lower temperature.

5.4.2 *Methane*

Methane does not adsorb directly onto the Ir(111) surface, but rather it immediately dehydrogenates to CH_3 upon reaching the surface. In order to determine the energy barrier for this process a NEB calculation was performed. The energy profile for this reaction is shown in Fig. 5.13.

The calculation shows that there is a 0.33 eV energy barrier for the process $CH_4(g) \rightarrow CH_3(ads) + H(ads)$. This is similar to the 0.3 eV value reported in [10]. Based on a partial pressure of $P_i = 1 \times 10^{-4}$ mbar (as in [11]), and an adsorption site area of $A_{site} = 2\sqrt{3}D$, the rate of methane adsorption is $2.06 \times 10^2 T_{gas}^{-0.5} \exp(-0.33/k_B T)\ \mathrm{s}^{-1}$.

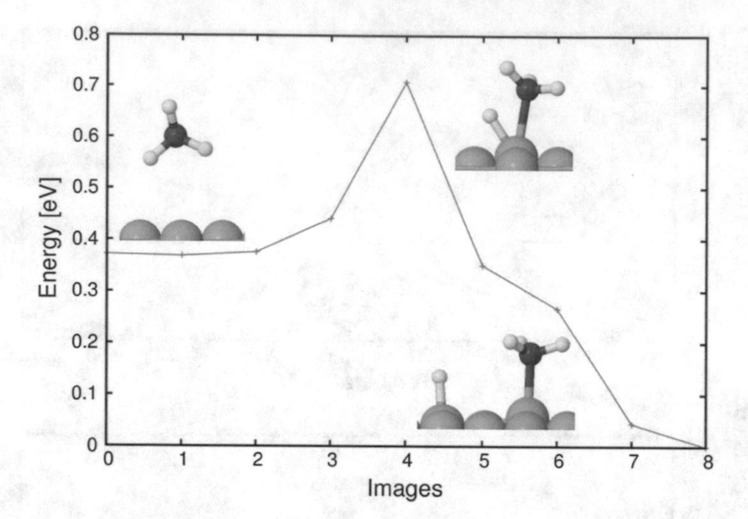

Fig. 5.13 The energy profile for methane adsorption onto the Ir(111) surface. It dehydrogenates upon adsorption to form CH_4 (g) → CH_3 (ads) + H (ads)

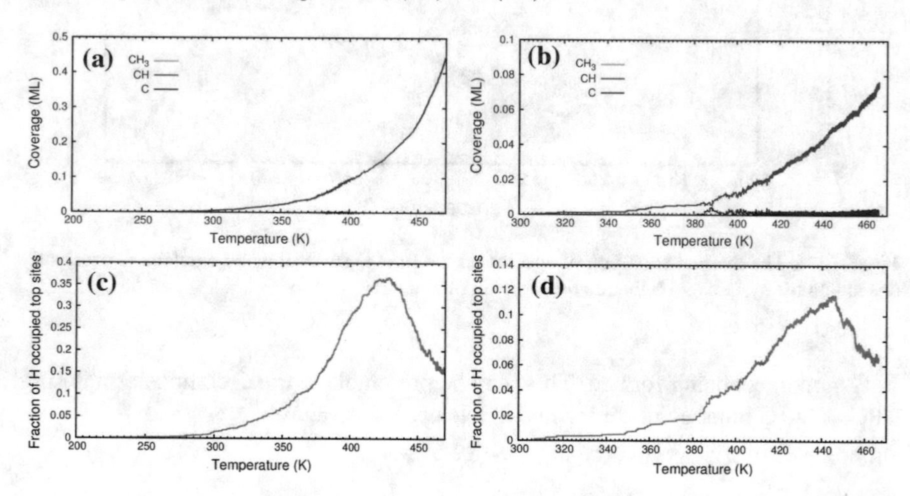

Fig. 5.14 The thermal evolution of species on the Ir(111) surface, while methane is constantly dosed into the system at a partial pressure of **a** 1×10^{-4} mbar and **b** 1×10^{-5} mbar. The coverage of H atoms on the surface relative to the number of free *top* sites at 1×10^{-4} mbar and 1×10^{-5} mbar is shown in **c** and **d**

The decomposition of methane as determined from the kMC simulations at a partial pressure of 1×10^{-4} and 1×10^{-5} mbar is shown in Fig. 5.14a, b. The overall mechanism involves the subsequent removal of H atoms to form C monomers. All the necessary barriers, apart from the barrier for the adsorption of methane, can be taken from the ethylene study discussed above. Initially methane adsorbs onto the surface by dehydrogenating CH_3. This then rapidly dehydrogenates to CH. Three H atoms are removed in quick succession since the energy barriers for the dehydrogenation of CH_3 and CH_2 are only 0.50 and 0.09 eV. As shown in Fig. 5.14a, b the temperature range in which methane adsorbs is dependent on the partial pressure of the gas. In both cases after methane adsorption CH remains on the surface for a relatively long time since any C monomers formed will rehydrogenate easily, as the energy barrier is small and there are a large amount of H atoms on the surface. H atoms are lost from the surface above 400 K, but depending on the partial pressure of methane, this H can be replaced by the absorption of methane. This adds four H atoms to the system for every C atom. With the lower partial pressure in Fig. 5.14b C monomers are formed more readily as there is less methane, and hence less H being added to the surface. This suggests that the decomposition of methane may favour higher partial pressures of methane gas initially, and then the deposition should be reduced (or stopped) in order to allow H to be removed from the system so that C monomers can form.

5.5 Conclusions

In this chapter kMC simulations of the thermal decomposition of hydrocarbons on the Ir(111) surface were performed. Details of the kMC code were described in detail in Chap. 4.

The main focus was the temperature programmed decomposition of ethylene on the Ir(111) surface, which was studied in Chap. 3 with the use of rate equations and XPS experiments. Using the same energy barriers as calculated in Chap. 3 the thermal decomposition process was simulated using the kMC method. Two surface coverages were considered: a low coverage simulation at 0.05 ML, and a high coverage simulation at 0.6 ML. The results at the high coverage were found to have qualitative agreement with the experimental results. By comparing the low and high coverage simulations the coverage effects incorporated with the kMC simulation were revealed: at high coverages more CHCH is produced since the reaction for the dehydrogenation of CH_2CH becomes more favourable at high coverages. This is supported by the experimental results in Fig. 3.7.

One failing of the kMC simulations is that the experimental coverages of species over the course of the temperature ramp were not reproduced. For example at low coverages the kMC simulations determine much more CH_3C than CHCH between 250 and 400 K, whereas the experimental results indicate that CHCH should be the dominating species in this temperature range. This discrepancy can be explained by small errors (5–10%) in the DFT calculated energy barriers, which may be due to the inherent errors in the DFT method or molecule interactions (at high coverages). As there are many competing reaction processes with similar energy barriers small changes in the barrier values can affect the kMC results significantly. By altering the barriers slightly, as shown in Table 5.1, kMC coverages which are in good agreement with the experimental results were recovered.

The thermal decomposition of ethylene at both high and low coverages was also modelled by solving rate equations (Fig. 5.9). The low coverage results (0.05 ML) were found to be almost identical to the kMC results at low coverage. However in the high coverage results there were differences between the kMC results and the rate equations. The rate equations predict a smaller temperature window for CHCH. This is a failing of the rate equations method, which does not include coverage effects. The kMC simulations, with the inclusion of coverage effects, find that more CHCH is produced at higher coverages, which as previously mentioned, is in agreement with the experimental findings.

The fixed temperature thermal decomposition of ethylene was also simulated for four different temperatures. Although the decomposition mechanism did not vary from the temperature ramping case, the results suggested that lower coverages might speed up the decomposition process slightly as some reactions may be slowed if there are not many unoccupied sites.

Overall the decomposition mechanism for ethylene on the Ir(111) was found to be:

$$\begin{aligned}
&CH_2CH_2 \rightarrow CH_2CH + H\\
&CH_2CH \longleftrightarrow CH_2C + H\\
&CH_2CH \rightarrow CHCH + H\\
&CH_2C + H \longleftrightarrow CH_3C\\
&CHCH \rightarrow CH + CH\\
&CH \rightarrow C + H
\end{aligned}$$

This is in agreement with what was found in Chap. 3. Based on the simulations performed in this chapter the limiting step in this process is the dehydrogenation of CH to form C monomers. This can only occur once the concentration of H atoms on the surface is low, otherwise C will prefer to rehydrogenate to CH. In these simulations H is removed from the surface by the formation of H_2 molecules, a process which only starts to occur above 400 K. Therefore for graphene growth from ethylene on the Ir(111) surface temperatures above 500 K are required in order to ensure that C monomers are formed.

The decomposition of ethylene on the Pt(111) surface was also simulated based on the energy barriers calculated in [1]. The overall mechanism was found to be the same as that on the Ir(111) surface, however the relative species coverages are heavily affected by the amount of H atoms available. The energy barrier for H_2(g) formation is 0.91 eV and between 200 and 400 K all the hydrogen produced from the dehydrogenation reactions is gradually lost to the gas phase. This means, that once most of the H has been lost CH_2C can no longer easily hydrogenate to CH_3C. Eventually, above 450 K CH_2C slowly transforms to CHCH, which breaks into two CH molecules, and then C is readily produced as most of the H has already been lost. Unlike the decomposition on Ir(111), depositing some H_2(g) may speed up the decomposition process on Pt(111) since it will assist the hydrogenation of CH_2C, which is required to form CHCH and continue the decomposition.

Chemical vapour deposition, where hydrocarbon molecules are supplied to the surface with a continuous flux as the system is heated was also investigated. The thermal decomposition of ethylene and methane were simulated. For ethylene CVD it was found that there was no major difference to the TPG decomposition mechanism. The main difference in species evolution was the slow increase in total species on the surface, due to the continuous ethylene deposition. C monomers were also found to begin forming at a slightly lower temperature when compared to TPG. This is possibly due to the presence of additional molecules from the deposition of ethylene throughout the temperature ramp.

When methane is deposited it adsorbs as CH_3 and then dehydrogenates to CH. At this stage the formation of C monomers is exactly the same processes as with ethylene deposition, requiring the removal of most of the H from the surface first. In both cases the decomposition process to form C monomers is limited by the large amount of H on the surface, which only starts to form H_2(g) above 400 K. If methane is continuously deposited onto the surface then the formation of C monomers may be limited by the additional hydrogen which will be added to the surface as methane dehydrogenates.

References

1. Y. Chen, D.G. Vlachos, Hydrogenation of ethylene and dehydrogenation and hydrogenolysis of ethane on Pt(111) and Pt(211): a density functional theory study. J. Phys. Chem. C **114**(11), 4973–4982 (2010)
2. Z. Li, P. Wu, C. Wang, X. Fan, W. Zhang, X. Zhai, C. Zeng, Z. Li, J. Yang, J. Hou, Low-temperature growth of graphene by chemical vapor deposition using solid and liquid carbon sources. ACS Nano **5**(4), 3385–3390 (2011)
3. H. Tetlow, J. Posthuma de Boer, I.J. Ford, D.D. Vvedensky, D. Curcio, L. Omiciuolo, S. Lizzit, A. Baraldi, L. Kantorovich, Ethylene decomposition on Ir(111): initial path to graphene formation. Phys. Chem. Chem. Phys. **18**, 27897–27909 (2016)
4. L. Kantorovich, *Quantum Theory of the Solid State: An Introduction* (Kluwer, 2004)
5. J. Coraux, A.T. N'Diaye, M. Engler, C. Busse, D. Wall, N. Buckanie, F.-J.M. zu. Heringdorf, R. van Gastel, B. Poelsema, T. Michely, Growth of graphene on Ir(111). New J. Phys. **11**(023006) (2009)
6. H. Tetlow, J. Posthuma de Boer, I.J. Ford, D.D. Vvedensky, J. Coraux, L. Kantorovich, Growth of epitaxial graphene: Theory and experiment. Phys. Rep. **542**(3), 195–295 (2014)
7. T. Fuhrmann, M. Kinne, B. Tränkenschuh, C. Papp, J.F. Zhu, R. Denecke, H.-P. Steinrück, Activated adsorption of methane on Pt(111) - an in situ XPS study. New J. Phys. **7**, 107–107 (2005)
8. H.A. Aleksandrov, L.V. Moskaleva, Z.-J. Zhao, D. Basaran, Z.-X. Chen, D. Mei, N. Rösch, Ethylene conversion to ethylidyne on Pd(111) and Pt(111): A first-principles-based kinetic Monte Carlo study. J. Catal. **285**(1), 187–195 (2012)
9. H.L. Abbott, I. Harrison, Dissociative chemisorption and energy transfer for methane on Ir(111). J. Phys. Chem. B **109**(20), 10371–10380 (2005)
10. G. Henkelman, H. Jónsson, Theoretical calculations of dissociative adsorption of CH_4 on an Ir(111) surface. Phys. Rev. Lett. **86**, 664–667 (2001)
11. D.C. Seets, C.T. Reeves, B.A. Ferguson, M.C. Wheeler, C.B. Mullins, Dissociative chemisorption of methane on Ir(111): evidence for direct and trapping-mediated mechanisms. J. Chem. Phys. **107**(23) (1997)

Chapter 6
Beginnings of Growth: Carbon Cluster Nucleation

In this chapter the nucleation of carbon clusters on the Ir(111) surface is studied. This process is important in the early stages of graphene growth, where the growth begins with the nucleation of small carbon clusters from carbon monomers, which then grow in size to form graphene islands.

To study the nucleation of carbon clusters the work of formation (or grand potential difference), $\Delta\phi(N)$ for forming a cluster of size N carbon atoms from the C monomer gas is considered. Unlike the zero temperature formation energy, which has been calculated in previous work on carbon clusters [1–3] this term is temperature dependent. This is far more relevant when considering the high temperatures that are used in graphene growth experiments.

To begin, an expression for the work of formation, $\Delta\phi(N)$ will be derived for carbon clusters on the Ir(111) terrace based on classical nucleation theory. This will include the cluster vibrational free energy and entropic terms. Using DFT calculations the necessary components of $\Delta\phi(N)$ can be determined for a variety of clusters, including different structural types, with a size range of $N = 1 - 16$. From this the stability of different cluster types can be found, as well as the critical cluster size over a range of temperatures.

Finally a general understanding of the evolution of cluster growth up to $N = 16$ is predicted. The mechanism transforming between different cluster structures will be analysed using nudged elastic band calculations.

6.1 Classical Nucleation Theory

According to classical nucleation theory the rate of the nucleation can be found by determining the size of the critical cluster, which contains a particular number of carbon atoms N^* [4]. Forming this cluster requires overcoming a nucleation barrier which is determined from the maximum of the cluster work of formation. The work of formation (or grand potential difference), $\Delta\phi(N)$ for forming a cluster of size

H.A. Tetlow, *Theoretical Modeling of Epitaxial Graphene Growth on the Ir(111) Surface*, Springer Theses, DOI 10.1007/978-3-319-65972-5_6

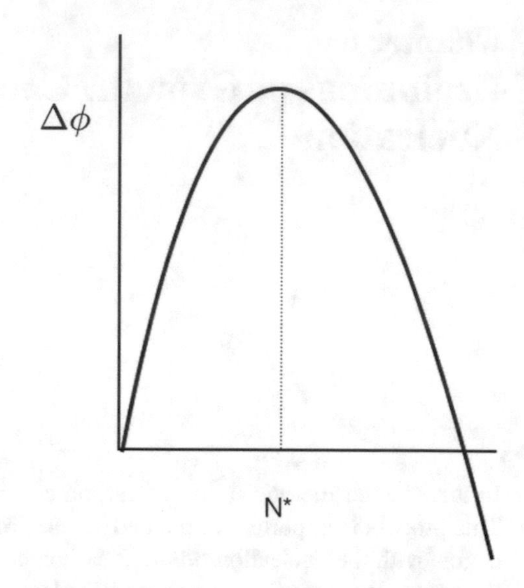

Fig. 6.1 The work of formation $\Delta\phi(N)$ as a function of cluster size N. The critical cluster size, N^*, given by the maximum of $\Delta\phi(N)$, corresponds to the size where monomer addition is favourable to removal and growth may become spontaneous

N is shown in Fig. 6.1. Once overcome the addition of atoms to a cluster is more favourable than their removal, and thus this sets the minimum size for stable clusters to form.

The appropriate quantity for studying the nucleation of carbon clusters is the work of formation, $\Delta\phi(N)$ for forming a cluster of size N carbon atoms from the C monomer gas. Before this can be calculated for each of the carbon clusters it must be derived.

6.1.1 Derivation of $\Delta\phi(N)$

Starting from classical nucleation theory it is possible to consider a stability of a cluster of N atoms in terms of its rate of formation, per unit volume, from a gas of monomers [4],

$$J = n_1 Z \beta_{N^*} \exp(-\Delta\phi(N)/k_B T) , \tag{6.1}$$

where n_1 is the monomer concentration, Z is the Zeldovich factor, k_B is Boltzmann's constant and β_{N^*} is the rate of monomer attachment to a critical cluster size N^*. If $\phi(N)$ is the grand potential for the N atom cluster absorbed on the surface, then the difference $\Delta\phi(N) = \phi(N) - \phi(1)$ in the exponent in Eq. (6.1) corresponds to the grand potential of the cluster relative to a single carbon atom. The maximum of this quantity with respect to N gives the barrier for nucleation and the corresponding critical cluster size N^*.

For an N atom cluster the potential $\phi(N)$ can be expressed in terms of the cluster free energy $F(N)$ and the chemical potential of the monomer gas μ in the following way:

$$\phi(N) = F(N) - N\mu. \tag{6.2}$$

Here the free energy $F(N) = -k_B T \ln Z(N)$, where $Z(N)$ is the cluster partition function which for solid-like clusters can take the form of

$$Z(N) = N_{\text{sites}} N_{\text{rot}} \exp(-U(N)/k_B T) Z^{\text{vib}}(N). \tag{6.3}$$

This expression contains entropic multiplicity terms related to the number of locations the cluster can occupy on a finite substrate of N_{sites} sites, and the number of rotational variants, N_{rot}, a cluster can have at the same lattice site (which will depend on its shape). It is assumed that there is a sufficiently low concentration of clusters on the surface so that one may neglect the interaction between different clusters. Then, the exponential Boltzmann factor contains the zero temperature energy $U(N)$ of a single cluster, while the last factor, $Z^{\text{vib}}(N)$, accommodates the appropriate vibrational contribution for the cluster resulting from the surface and a single cluster adsorbed on it. The energy can be written $U(N) = U_0 + \Delta U(N)$, where U_0 is the energy of the isolated surface and $\Delta U(N)$ contains the energy of an isolated cluster, interaction energy of the cluster with the surface as well as the relaxation energy of both the surface and the cluster.

Correspondingly, the free energy of the cluster on the surface is:

$$F(N) = -k_B T \ln(N_{\text{sites}} N_{\text{rot}}) + U_0 + \Delta U(N) + F^{\text{vib}}(N)\,, \tag{6.4}$$

where $F^{\text{vib}}(N) = -k_B T \ln Z^{\text{vib}}(N)$ is the vibrational contribution to the free energy. To calculate the latter, we note that the combined vibrational density of states (DOS) of the cluster and surface system $D(\omega)$ can be expressed as,

$$D(\omega) = D_0(\omega) + \Delta D_N(\omega)\,, \tag{6.5}$$

where $D_0(\omega)$ is the DOS of the isolated surface and

$$\Delta D_N(\omega) = \sum_{\lambda \in S + C_N} \delta(\omega - \omega_\lambda) - \sum_{\lambda \in S_0} \delta(\omega - \omega_\lambda)$$

is the change in the total DOS due to the adsorbed C_N cluster. The first term in the last formula contains all vibrational modes of the cluster and surface system $C_N + S$, while the second contains modes associated only with the isolated surface, S_0. Hence,

$$F^{\text{vib}}(N) = F_0^{\text{vib}} + \Delta F^{\text{vib}}\,,$$

where F_0^{vib} is the vibrational free energy of the isolated surface, and

$$\Delta F^{\text{vib}}(N) = -k_B T \int \Delta D_N(\omega) \ln Z^{\text{vib}}(\omega) d\omega = \int \Delta D_N(\omega) F^{\text{vib}}(\omega) d\omega$$
$$= \sum_{\lambda \in S+C_N} F^{vib}(\omega_\lambda) - \sum_{\lambda \in S_0} F^{vib}(\omega_\lambda) \tag{6.6}$$

is the free energy change due to the adsorbed cluster. The free energy due to a single vibrational mode of frequency ω can be expressed within the quasi-harmonic approximation as

$$F^{\text{vib}}(\omega) = -k_B T \, \ln Z^{\text{vib}}(\omega) = \frac{1}{2}\hbar\omega + k_B T \, \ln\left(1 - \mathrm{e}^{-\hbar\omega/k_B T}\right). \tag{6.7}$$

Combining all the expressions given above, we obtain for the grand potential of the cluster on the surface an expression:

$$\phi(N) = -k_B T \, \ln(N_{\text{sites}} N_{\text{rot}}) + \left(U_0 + F_0^{\text{vib}}\right) + \Delta U(N) + \Delta F^{\text{vib}}(N) - \mu N \, . \tag{6.8}$$

To calculate the required difference of the grand potentials, $\Delta\phi(N) = \phi(N) - \phi(1)$, the quantity $\phi(1)$ is also required, which is found by setting $N = 1$ in the expression above. The $U_0 + F_0^{\text{vib}}$ term is cancelled out in the difference.

The chemical potential of the monomer gas μ of N carbon atoms on the surface at temperature T also needs to be calculated. For N carbon atoms distributed on N_{sites} surface sites, there are $N_{\text{sites}}!/\left[N!\,(N_{\text{sites}} - N)!\right]$ possibilities. Assuming that $N \ll N_{\text{sites}}$ (the limit of small concentration), interaction between carbon atoms may be neglected. Then the total energy of the system $U^{\text{C}}(N) = U_0 + N\Delta U(1)$ is additive, where $\Delta U(1)$ is the energy of a single adsorbed C atom on the surface calculated relative to the energy U_0 of the isolated surface. Similarly, the vibrational free energy $F_{\text{C}}^{\text{vib}} = F_0^{\text{vib}} + N\Delta F^{\text{vib}}(1)$ is also additive, with $\Delta F^{\text{vib}}(1)$ being the change of the free energy of the surface and a single C atom due to its adsorption. It is given by the expression analogous to Eq. (6.6). Therefore, repeating the arguments employed in deriving Eq. (6.8), the free energy of N non-interacting carbon atoms on the surface (the monomer gas) is

$$F_m(N) = -k_B T \ln \frac{N_{\text{sites}}!}{N!(N_{\text{sites}} - N)!} + \left(U_0 + F_0^{\text{vib}}\right) + N\Delta U(1) + N\Delta F^{\text{vib}}(1) \, . \tag{6.9}$$

The chemical potential of the monomer gas is then obtained from its definition as

$$\mu = \left(\frac{\partial F_m}{\partial N}\right)_T = \Delta U(1) + \Delta F^{\text{vib}}(1) - k_B T \, \ln \frac{1-\theta}{\theta} \, , \tag{6.10}$$

where $\theta = N/N_{\text{sites}}$ is the monomer coverage and Stirling's approximation was used.

Combining the expression (6.8) for the grand potential and that for the chemical potential of the free monomer gas of carbon atoms, Eq. (6.10), the expression for the grand potential difference is finally obtained:

$$\Delta\phi(N) = [\Delta U(N) - N\Delta U(1)] + \left[\Delta F^{\text{vib}}(N) - N\Delta F^{\text{vib}}(1)\right] - k_B T \ln N_{\text{rot}} + (N-1)k_B T \ln \frac{1-\theta}{\theta}. \quad (6.11)$$

Note that $\Delta\phi(N)$ depends not only on N, but also on the monomer coverage, θ, and the temperature, T.

6.2 Carbon Clusters

The first step in determining the work of formation for the carbon clusters is to find suitable cluster structures, which are energetically favourable. For small clusters this may be simple, since there are only a few possible structures, however as the clusters grow in size there are many options for producing stable structures. To proceed the clusters are divided into different types (chains, arches, rings, compact, and domes). Images of the most energetically stable structures of these different cluster types are shown in Fig. 6.2 for a variety of sizes N.

For carbon monomers the most stable position is in the hcp site. For the C_2 and C_3 structures only the chain structures may be formed, where each atom sits in neighbouring hollow sites. For larger clusters multiple types of cluster are possible. As well as the flat linear chain structures, arching clusters may also be formed where the linear chain of atoms bend over the surface. Often only the atoms at the ends of the arch interact with the surface, while the inner atoms bend upwards away from the surface.

Several types of compact structures are also possible. Stable compact clusters can be formed by arranging the C atoms in alternating top and hollow sites, referred to as top-hollow (TH) compact clusters. These clusters are slightly dome-like, where the inner atoms interact less strongly with the surface than the outer atoms. A different type of compact cluster can also be formed by arranging the C atoms in a closed ring around the Ir surface atoms. For clusters containing 10 atoms or more dome-like (DL) clusters can be formed by creating pentagonal and hexagonal rings of C atoms, which usually enclose Ir atoms. Some of these have been proposed previously in the literature [6, 7].

6.2.1 Rotational Multiplicity

The work of formation of a cluster (Eq. (6.11)), contains an entropic term that is dependent on the cluster's rotational multiplicity, N_{rot}. N_{rot} can be found by

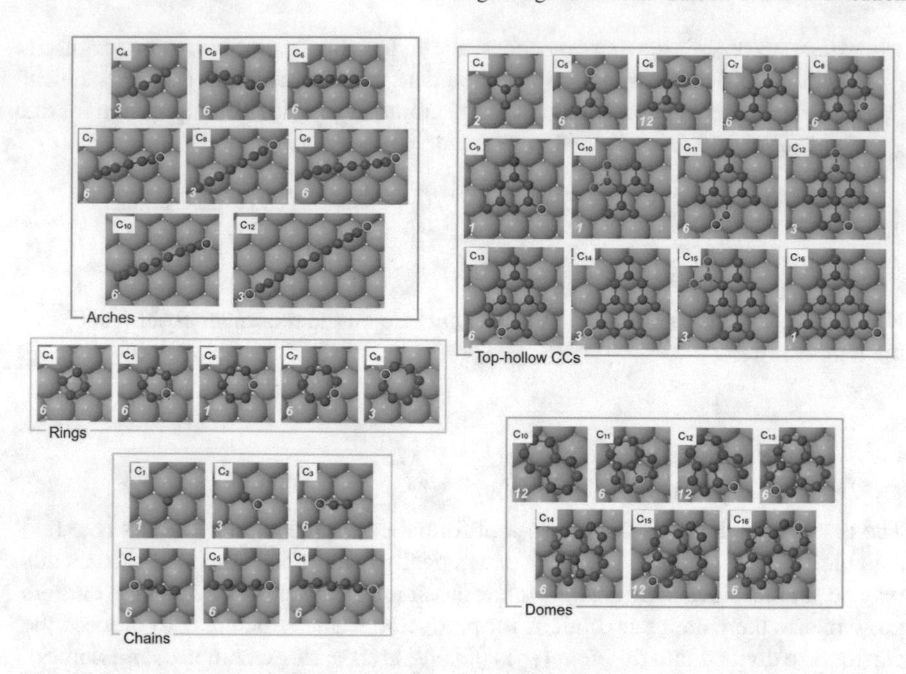

Fig. 6.2 The most energetically stable clusters for each cluster type for various N. The number of carbon atoms in each cluster is shown in the *upper-left* corner of each figure, and the rotational multiplicities N_{rot} are shown in the *lower-left* corner. The clusters are grouped in their five different types: arches, rings, top-hollow, chains and domes. Highlighted *yellow* atoms indicate how the given clusters could be formed by adding a C atom from the previous one in the same box (see text). The highlighted *green* atoms show the additional atoms required to jump between the most stable top-hollow cluster structures (C_7, C_{10}, C_{12}, C_{15}). Reprinted from [5], with the permission of AIP Publishing

examining the rotational degrees of freedom for each cluster on the hexagonal fcc(111) surface. For example the C_9 compact cluster shown in Fig. 6.2, has 2 rotational degrees of freedom due to its 3-fold symmetry on the fcc(111) surface. For each cluster in Fig. 6.2 N_{rot} is written in the bottom-left corner.

6.3 Zero-Temperature Formation Energy

The work of formation at 0 K can be calculated by neglecting all the temperature dependent terms in Eq. (6.11). The 0 K work of formation (commonly referred to as the formation energy) becomes

$$E_f(N)_{T=0} = \Delta U(N) - N\Delta U(1). \tag{6.12}$$

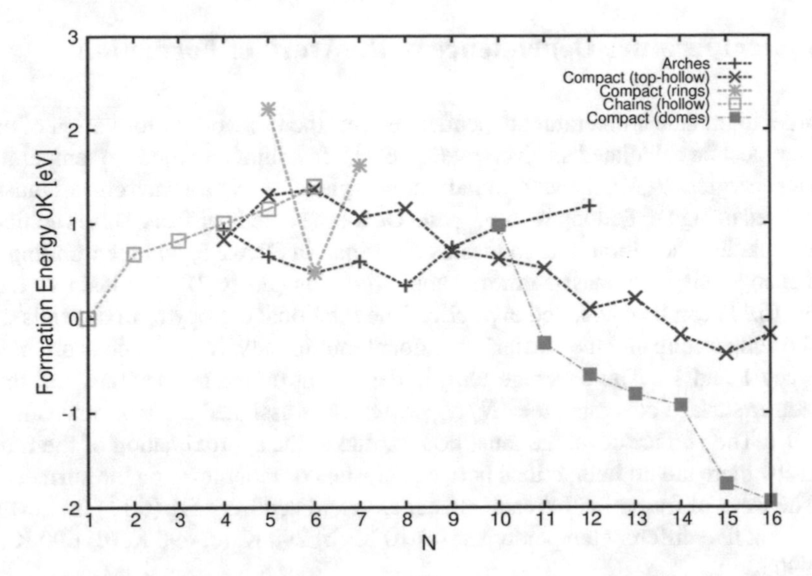

Fig. 6.3 The 0 K formation energy of the various clusters as calculated from Eq. (6.12). Reprinted from [5], with the permission of AIP Publishing

This quantity can be used to compare the relative stability of the clusters at 0 K. The formation energy (FE) of all the clusters shown in Fig. 6.2 is presented in Fig. 6.3. The different cluster types are indicated by the colours. Linear chains and arches are shown in yellow and red respectively, whereas the compact clusters: top-hollow (TH) compacts, rings, and dome-like (DL) clusters are shown in blue, green and purple.

Starting from C_4 clusters, it is noticeable that the chain, arch and TH-compact clusters have a similar formation energy of around 1 eV. Of these, the TH-compact structure has the lowest formation energy. Between $N = 5 - 9$ arches are the most stable cluster type. Exceptions to this are C_6, where the hexagonal ring cluster has the same formation energy as the arch, and C_9 where the TH-compact cluster energy matches that of the C_9-arch structure. This is a result of the symmetry of these particular clusters on the fcc (111) surface. Above C_{10} compact clusters become preferable to the linear arching clusters. Broadly the DL clusters are more stable than the TH-compact clusters; however, for C_{10} the TH-compact structures are more stable.

These results suggest that linear (arching) clusters are more stable initially. However as the cluster size increases compact clusters become stable. This is likely due to the fact that at large sizes the structures become stabilised by the outer atoms, causing the inner atoms to rise up from the surface. This leads both the TH-compact clusters and the DL clusters to have a dome-like shape. It is highly possible that at larger N these hexagonal based structures should continue to prevail, leading to graphene formation. However it is questionable to make conclusions about the carbon clusters in graphene growth without including temperature effects into the cluster work of formation.

6.4 Temperature Dependence of the Work of Formation

In order to include temperature dependent effects the full work of formation of each cluster must be calculated as given by Eq. (6.11). In addition to the zero temperature cluster energies, $U(N)$, the vibrational free energies $F_{vib}(N)$ for each cluster must be calculated using DFT, along with F^0_{vib} and U_0 for the bare Ir surface. This is achieved by calculating the vibrational modes as described in Sect. 2.5, and then finding the free energy using the quasi-harmonic approximation (Eq. (6.7)). The last two terms of Eq. (6.11) can be calculated explicitly. The rotational entropy term depends only on the temperature and the cluster's rotational multiplicity, N_{rot}, which is an integer between 1 and 12. The coverage term is dependent on the temperature and the C monomer surface coverage, $\theta = N/N_{sites}$, which is assigned the arbitrary value of $\theta = 0.1$. The surface coverage must be low, due to the approximation of the model whereby there are no interactions between clusters or monomers on the surface.

The work of formation for each cluster as calculated from Eq. (6.11) is shown in Fig. 6.4 at five different temperatures: (a) 10 K, (b) 290 K, (c) 490 K, (d) 690 K and (e) 990 K.

At 10 K the work of formation of the various clusters appears very similar to the formation energies at 0 K (Fig. 6.3). As the temperature is increased it is noticeable that work of formation increases, especially for larger N. At $N = 16$ $\Delta\phi(N)$ increases by as much as 2 eV from 10–990 K. The temperature also changes the critical cluster size N^* and the nucleation barrier $\Delta\phi(N^*)$. Starting at 10 K it can be seen that there are two values for N where there is a noticeable nucleation barrier, which are almost of the same height. The first of these occurs at $N = 4$, which coincides with a transition in the most stable cluster structure type between chains and arches. The other barrier occurs at $N = 9$, where compact structures start to become more stable. Between 10 and 490 K both these barriers are similar in height, and therefore should both be equally limiting to the cluster growth progression. The arch-shaped clusters for the values of N lying between these two barriers should be stable to both the addition and removal of C monomers unless the second nucleation barrier is overcome. As the temperature is increased $\Delta\phi(0)$ increases whereas $\Delta\phi(4)$ remains approximately the same. Between these two values there is a minimum for the arching clusters at $N = 5 - 6$. This suggests that these clusters should be somewhat stable compared to the others. At 990 K the second barrier increases to ~ 2 eV and N^* increases to 10. At high temperatures clusters larger than $N = 12$ should be expected to be stable. Furthermore at $N = 15 - 16$ there is a rapid increase in the stability of the DL clusters.

Across all temperature ranges linear structures are initially more stable at smaller N. Below $N = 4$ chain clusters are more stable whereas between $N = 4$ and 9 arching clusters are more stable. Above $N = 10$ compact clusters become more stable, especially DL clusters. This is in agreement with the results of others, which suggested that for $N = 1 - 10$ non-compact clusters should be more stable than compact clusters [1, 2]. The fact that the critical cluster sizes N^* coincide with these values of N suggests that the cluster growth may be limited by the ability for clusters to transform their structure type. This is discussed in Sect. 6.6.

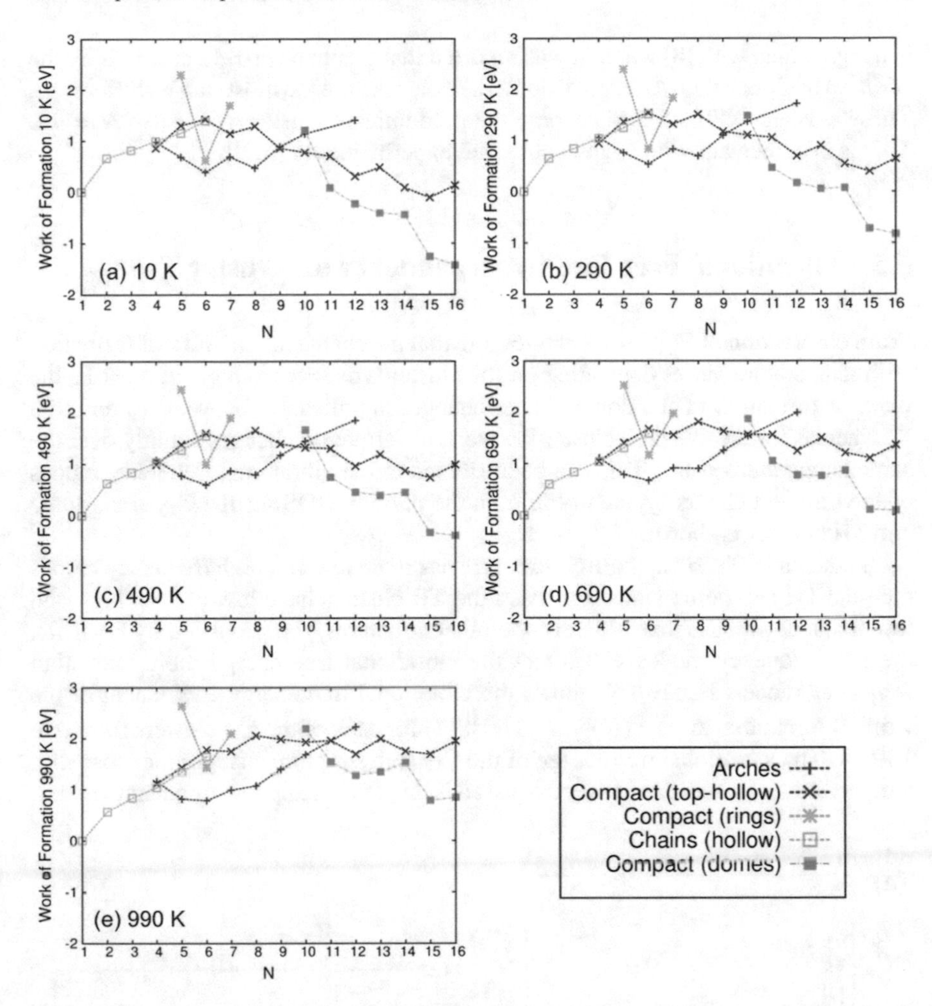

Fig. 6.4 The work of formation (the grand potential difference, $\Delta\phi(N)$) of various carbon clusters at different temperatures: **a** 10 K, **b** 290 K, **c** 490 K, **d** 690 K, **e** 990 K. The work of formation is shown relative to that for the single carbon atom adsorbed on the surface. Reprinted from [5], with the permission of AIP Publishing

Based on these results cluster formation prior to graphene growth can be understood. The presence of two critical cluster sizes in the work of formation suggests that between these values of N^* $(4 < N < 9)$ arching clusters may be stable on the surface. However for other cluster types this is not the case. Only once the second nucleation barrier is overcome cluster growth will become favourable. As this barrier increases with temperature it may be difficult to form larger clusters via monomer attachment. Instead it may be preferable for two small stable clusters (for example with $N = 5 - 6$) to coalesce to form a larger cluster. The stability of these small clusters suggests that they may contribute directly to the graphene growth front, which is in agreement with [8] where it was reported that graphene growth proceeds by the addition of C_5 clusters. At large N dome-like clusters are the most stable cluster type. These clusters will therefore become the predominant cluster type above $N = 12$. This is in agreement with what is observed experimentally [6, 9].

6.5 Vibrational Free Energy Dependence on Cluster Type

From observation of Fig. 6.4 it is noticeable that the change in the work of formation with temperature varies depending on the cluster type. For example at $N = 12$ the work of formation of the dome cluster changes significantly between 10 and 990 K, whereas for the C_{12} arch cluster the work of formation changes slightly over the same temperature range. This can be attributed to the vibrational frequency modes of the different cluster types. In Fig. 6.5a the phonon DOS for the C_{12} arch, dome and TH clusters is shown.

As shown in Fig. 6.5a, the arch structure has more low and high frequency modes than the TH and dome structures. Also the TH clusters have less of both high and low frequency modes than either of the other structures. As calculated by Eq. (6.7), the low frequency modes will affect the vibrational free energy more than high frequency modes. Figure 6.5b shows the vibrational free energy component of the work of formation, $\Delta F^{\text{vib}}(N) - N\Delta F^{\text{vib}}(1)$ for each of the C_{12} clusters from 0 to 1000 K. The vibrational free energy of the C_{12} arch structure changes the most with temperature, which as previously discussed is due to its many low frequency modes.

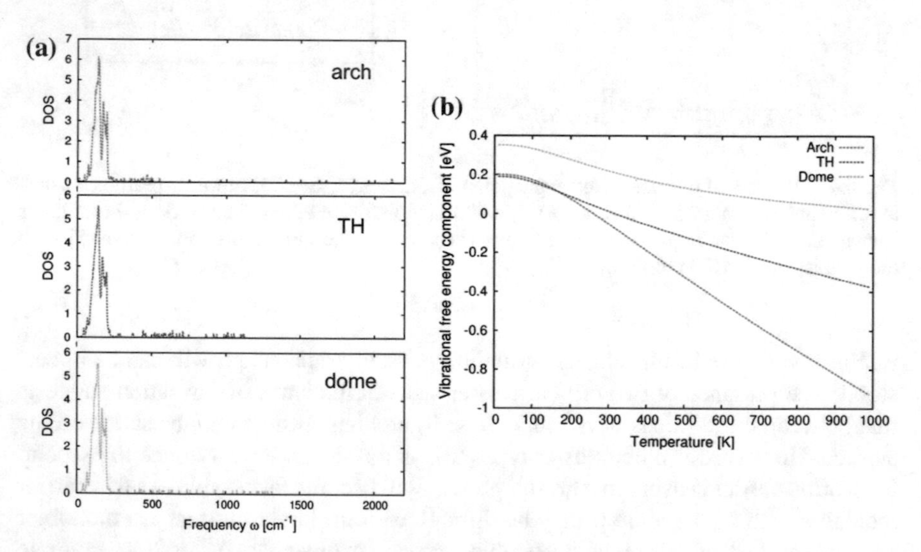

Fig. 6.5 **a** The phonon DOS and **b** the vibrational free energy component of the work of formation for the C_{12} arch, dome and TH clusters. Reprinted from [5], with the permission of AIP Publishing

In Fig. 6.4 this negative vibrational component is offset by the positive coverage term in the work of formation (Eq. (6.11)), and hence work of formation of the arch structure does not vary significantly with temperature. For the dome and TH clusters however, the vibrational free energy over all temperature ranges is positive or only slightly negative, resulting in a greater change in their work of formation as a function temperature. For the dome structure this effect is more pronounced since at higher temperatures its vibrational free energy decreases less. From these results it should be expected that there will be a similar trend for different sized clusters as the phonon DOS should depend on the structure of the cluster. Dome structures should on average have less low frequency modes, and hence will have a more positive vibrational free energy component in the work of formation. This will effect their work of formation, causing it to be less negative than other cluster types at high temperatures.

6.6 Cluster Isomerisation During Growth

The carbon clusters are grouped into their different types based on their structure. Because of the vast differences between different structure types it is likely that clusters cannot reconstruct from one type to another. This means that during growth if clusters are to grow by the addition of C monomers they should stay within the same cluster group. However, as noted from the zero-temperature formation energy results, different cluster types are stable over different size ranges. Therefore if compact clusters are to become the dominant cluster type on the surface at larger N then the linear arching clusters must either isomerise into compact clusters, or gradually decompose to smaller clusters that may then have the chance to regrow into a compact structure. The same logic can also be applied to the transition from TH-compact clusters to the DL clusters.

To investigate the transformation between cluster types the energy barriers for reconstructing clusters into different types are found using the NEB method. Two cluster reconstructions are studied, one from an C_7 arch cluster to an C_8 TH-compact cluster, and the other from an C_{10} TH-compact cluster to an C_{11} DL cluster. These represent the important stages during the cluster growth where the most stable cluster type changes. The energy barriers for forming the C_4 chain, arch and TH-compact structures from the C_3 chain are also calculated. These competitive processes are important at the onset of cluster growth since the process with the lowest energy barrier will direct the growth towards that particular cluster type.

The initial, final and transition states, along with the calculated energy profile for each of the reconstructions are shown in Fig. 6.6 for (a) the arch-compact and (b) the TH-DL reconstruction. For the arch-compact reconstruction a C atom is added to the centre of the arch, which then allows the inner atoms to connect to the surface and then flatten to form the compact structure. The energy barrier for this process is 1.07 eV, while the reverse barrier is 1.41 eV. For the TH-DL reconstruction a C atom is added to the TH-compact cluster to complete a hexagonal ring. The cluster

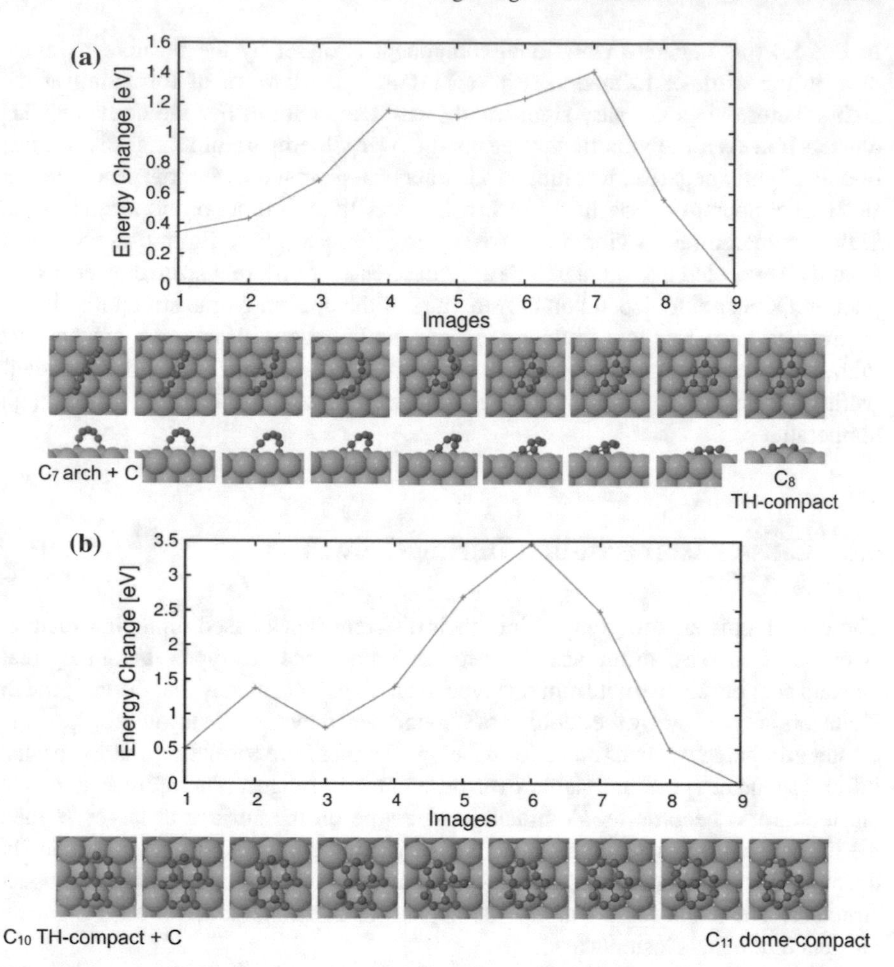

Fig. 6.6 The initial, final and intermediate states, along with the calculated energy profile, are shown for the reconstruction of **a** C arch cluster with an additional C atom into the C_8 TH cluster, and **b** C_{10} TH cluster with an additional C atom into the C_{11} dome. Reprinted from [5], with the permission of AIP Publishing

then rotates as the dangling C atoms move to close the ring and form pentagons. The forward process energy barrier is around 3 eV, and the reverse process barrier is around 3.45 eV.

The barriers for reconstructing the clusters between types are both large. This suggests that transforming between different types may be unlikely over the size range examined and the temperature considered. This is particularly the case for the TH-DL reconstruction where the barrier is around 3 eV since the cluster needs to rotate, which causes the breaking and formation of many bonds. If clusters of a particular type cannot transform into a lower energy type then it is possible that they will grow until they reach a size where they become too unstable, and then break apart.

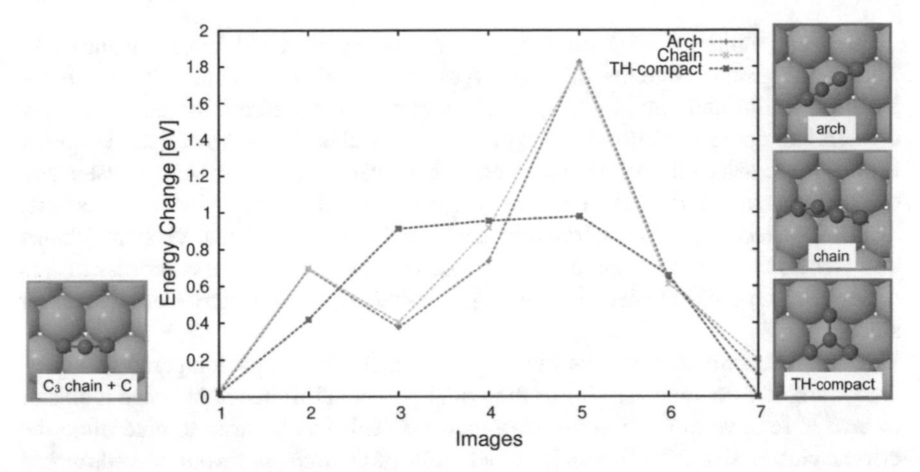

Fig. 6.7 The initial and final structures in the transformations of the C_3 chain into either C_4 chain, arch or TH clusters. Reprinted from [5], with the permission of AIP Publishing

There are many possible processes whereby different clusters can grow, reconstruct and decompose, which makes a complete study almost impossible. However based on the results for the work of formation, and the experimental work of others [6, 9] it is deduced that dome-like clusters are the dominating cluster type in the early stages of graphene growth. This suggests that these clusters must form by some other mechanism than addition of monomers, possibly by the coalescence of small stable clusters, such as the C_5–C_6 clusters indicated by the work of formation results.

For the C_3 to C_4 reactions the energy profiles are shown in Fig. 6.7. Forming the arch and chain structure from the C_3 cluster requires overcoming an energy barrier of around 1.8 eV. For forming the TH-compact cluster the barrier is 0.96 eV. All of the processes have a similar barrier for their reverse reactions to remove the C atom. For the C_3–C_4 reactions formation of the TH-compact cluster requires overcoming the smallest energy barrier. This cluster will likely be formed in preference to the chain and arch structures in the early stages of cluster growth. If cluster reconstruction requires overcoming large energy barriers then the compact clusters will continue to grow into larger compact clusters. This will lead to them becoming the predominant cluster type despite having a formation energy that is less favourable compared to the arching clusters.

6.7 Conclusions

In this chapter the nucleation of carbon clusters on the Ir(111) surface in the early stages of graphene growth was studied. The temperature dependent work of formation $\Delta\phi(N)$ was calculated for a variety of different cluster types with sizes ranging from $N = 1 - 16$.

To begin, multiple carbon cluster structures were considered for each value of N. These were grouped into five different types based on their structure. Linear clusters included chains and arches. Compact clusters were grouped into rings, top-hollow clusters and dome-like clusters. The lowest energy cluster of each of these types for each N were selected. The zero temperature formation energy for each cluster was calculated as in Eq. (6.12), and used to determine the relative stability of the clusters. This suggested that for clusters sized between $N = 1 - 10$ linear clusters (chains and arches) are the most stable cluster type, whereas for $N > 10$ compact clusters (primarily dome-like clusters) become more stable. This was in agreement with other studies [1, 2].

The novel component of this study was the calculation of the temperature dependent work of formation $\Delta\phi(N)$ or the grand potential difference between a cluster of size N relative to an N atom monomer gas. This can be used to determine the critical cluster size N^* for which the addition of C atoms is favourable compared to their removal, and the corresponding nucleation barrier $\Delta\phi(N^*)$. The work of formation was devised from scratch based on classical nucleation theory. In addition to the zero temperature cluster energy $U(N)$, this also includes terms depending on the vibrational free energy, the rotational degrees of freedom and the surface coverage. Correspondingly, the vibrational free energy was calculated for each cluster using DFT.

The results for the work of formation over a range of temperatures demonstrated the importance of temperature effects. As the temperature is increased the work of formation increases, especially for larger clusters. The temperature changes also affect the critical cluster size and nucleation barrier. Between 10 and 490 K there are two critical cluster sizes at $N = 4$ and $N = 10$, both with similar nucleation barrier heights. This results in a range of arching clusters between these values of N^* which should be stable on the surface. The minimum in this range is located between $N = 5 - 6$, and hence these clusters in particular should exist on the surface for long times, and may contribute directly to the graphene growth front, as suggested in [8]. As the temperature is increased the second barrier increases. This suggests that cluster growth by monomer attachment is unfavourable at high temperatures, and instead it may be more likely for smaller stable clusters to coalesce to form larger clusters. At large N domes are found to be the predominant cluster type.

Overall the calculated formation energy and the work of formation demonstrates that there are noticeable changes in the most stable cluster type over the range of N studied. This suggested that clusters will have to transform between the different structure types in order to stabilise their energy. To investigate this the energy barriers for the reconstruction between different cluster types was calculated using NEB calculations. Based on the energy barriers it was found that for adding a C atom to a C_3 chain the most preferable C_4 cluster type to form is the C_4 top-hollow cluster. Once formed it will then be more likely for larger top-hollow clusters to grow from it. For the reconstruction between the C_7 arch to the C_8 top-hollow cluster a 1 eV barrier was calculated. This would be easily overcome given the high temperatures used on graphene growth. However for the C_{10} compact to C_{11} dome-like reconstruction the barrier was 3 eV, due to the many bonds that need to be broken and reformed during the reconstruction. This suggests that a transformation from top-hollow clusters is unlikely. Instead domes may be formed from the coalescence of the long lived C_5 to C_6 clusters.

References

1. C. Herbig, E.H. Åhlgren, W. Jolie, C. Busse, J. Kotakoski, A.V. Krasheninnikov, T. Michely, Interfacial carbon nanoplatelet formation by ion irradiation of graphene on iridium(111). ACS Nano **8**(12), 12208–12218 (2014)
2. P. Wu, H. Jiang, W. Zhang, Z. Li, Z. Hou, J. Yang, Lattice mismatch induced nonlinear growth of graphene. J. Am. Chem. Soc. **134**, 6045–6051 (2012)
3. Q. Yuan, J. Gao, H. Shu, J. Zhao, X. Chen, F. Ding, Magic carbon clusters in the chemical vapor deposition growth of graphene. J. Am. Chem. Soc. **134**, 2970 (2012)
4. I.J. Ford, Nucleation theorems, the statistical mechanics of molecular clusters, and a revision of classical nucleation theory. Phys. Rev. E **56**, 5615–5629 (1997)
5. H. Tetlow, I.J. Ford, L. Kantorovich, A free energy study of carbon clusters on ir(111): Precursors to graphene growth. J. Chem. Phys. **146**(4), 044702 (2017)
6. Y. Cui, Q. Fu, H. Zhang, X. Bao, Formation of identical-size graphene nanoclusters on Ru(0001). Chem. Commun. **47**, 1470–1472 (2011)
7. P. Lacovig, M. Pozzo, D. Alfè, P. Vilmercati, A. Baraldi, S. Lizzit, Growth of dome-shaped carbon nanoislands on Ir(111): the intermediate between carbidic clusters and quasi-free-standing graphene. Phys. Rev. Lett. **103**(166101) (2009)
8. E. Loginova, N.C. Bartelt, P.J. Feibelman, K.F. McCarty, Evidence for growth by C cluster attachment. New J. Phys. **10**(093026) (2008)
9. B. Wang, X. Ma, M. Caffio, R. Schaub, W.-X. Li, Size-selective carbon nanoclusters as precursors to the growth of epitaxial graphene. Nano Lett. **11**, 424–430 (2011)

Chapter 7
Removing Defects: Healing Single Vacancy Defects

During epitaxial graphene growth vacancy defects may be formed in the graphene structure. These defects are undesirable since they affect the electronic properties of the graphene and reduce the carrier mobility [1]. However it was shown that it is possible for vacancy defects to heal once they are produced in the case of graphene grown on the Ir(111) surface, if the system is dosed with ethylene molecules as in CVD growth [2]. In this chapter the healing of graphene single vacancy defects via the deposition of ethylene molecules will be investigated with the use of molecular dynamics simulations and nudged elastic band calculations.

7.1 Theoretical Method Outline

The graphene/iridium system forms a moiré superstructure due to the lattice mismatch. As a result of this there are six different stable single vacancy structures. To begin, the relaxed geometries of the most energetically favourable of these is determined. The deposition of ethylene molecules, with their motion directed towards the defect site is simulated using molecular dynamics. To control the temperature of the system the NVT ensemble is used, with the Langevin thermostat. The damping parameter for this is verified by calculation of the phonon density of states based on the system dynamics. Multiple MD ethylene deposition simulations are performed. A variety of different initial conditions are attempted, such as the incident position of the ethylene molecule with respect to the defect site and initial molecule velocity. Finally, based on the MD simulations, a step by step defect healing mechanism is investigated, based on using NEB calculations to determine the energy barriers for each process.

H.A. Tetlow, *Theoretical Modeling of Epitaxial Graphene Growth on the Ir(111) Surface*, Springer Theses, DOI 10.1007/978-3-319-65972-5_7

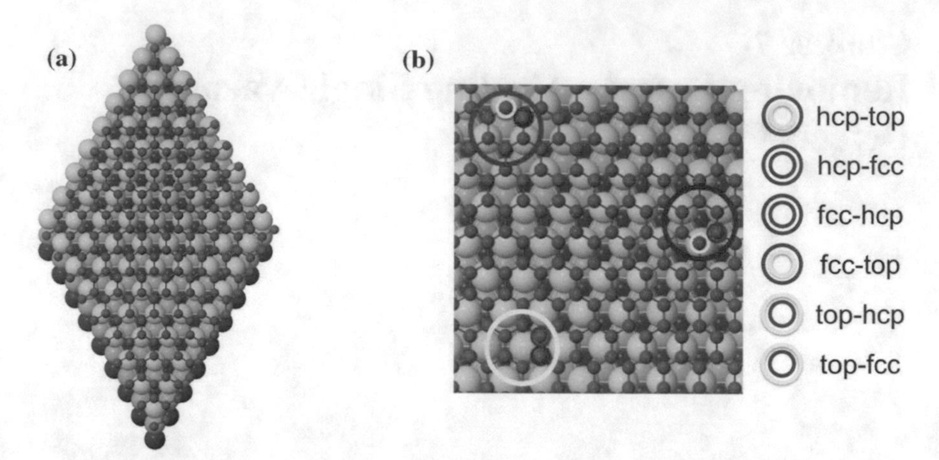

Fig. 7.1 **a** The graphene/iridium unit cell consisting of $(8 \times 8)/(9 \times 9)$ Gr/Ir units. **b** The high symmetry regions of the unit cell where single vacancy defects are stable as based on [3]. The *large circles* show the region, which is defined by the Ir site that the C-ring encircles. This makes up the first half of the defect name. The *smaller circles* show the C atoms which are removed to form the defect. These are defined by the Ir site that the C atom is above, which makes up the second half of the defect name

7.2 Single Vacancy Defects

The unit cell of the graphene/iridium superstructure is shown in Fig. 7.1a. There are three high symmetry region across the superstructure where the graphene hexagonal units are centred on a particular Ir lattice site. The top, hcp and fcc regions are marked on Fig. 7.1b. Single vacancy defects formed in these regions are the most stable due to the way the Ir and C atoms are laterally aligned. For each of these regions two single vacancy structure types may be formed by removing the C atom from above the respective Ir sites. For example in the fcc region the fcc-top and the fcc-hcp defects are possible. In total six different single vacancies may be formed in these regions. These are shown in Fig. 7.2a–f.

The various single vacancy defect types, as shown in Fig. 7.2, are relaxed with DFT to find which of them is lowest energy structure. This was found to be the hcp-top defect, which is in agreement with [4]. Based on this, this defect should form more commonly and be more stable than the other defect types. Therefore healing of this defect will be attempted with the MD simulations.

7.3 Langevin Thermostat

The MD simulations of the healing of a single vacancy require projecting an ethylene molecule with a high velocity on to the defect site. This introduces a large amount of energy into the system which in the microcanonical thermostat (NVE) would

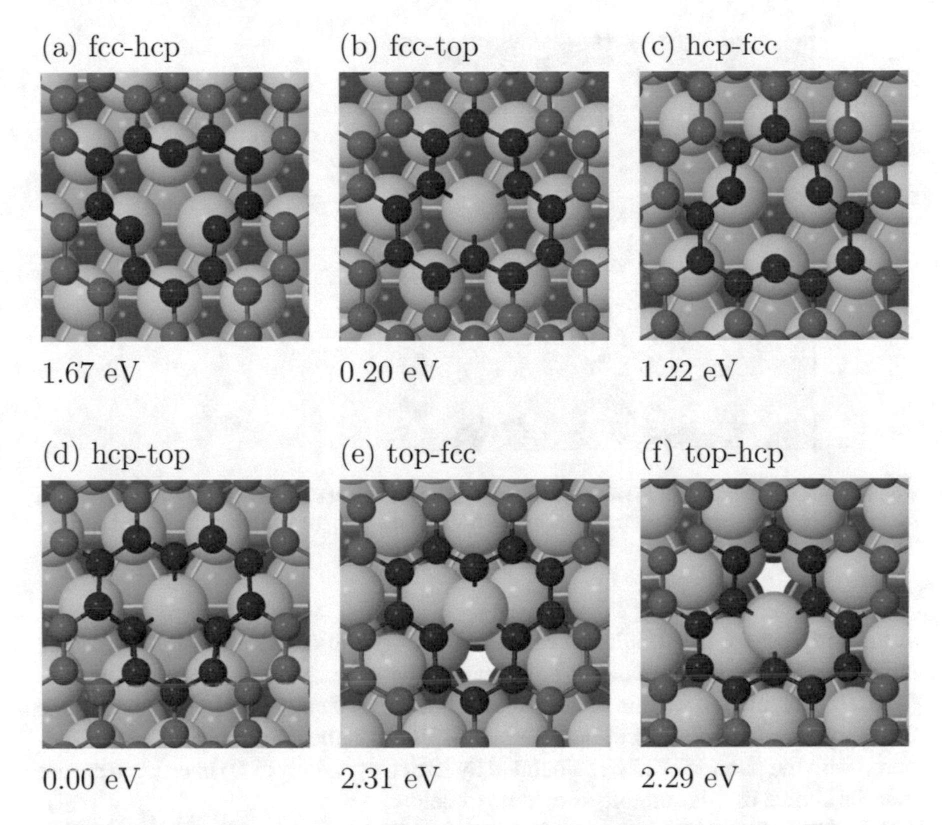

Fig. 7.2 The relaxed structures of the different SV defect types on the Ir(111) surface. The energies are given with reference to the hcp-top defect, which is the lowest energy structure. Note that gaps are shown in **e** and **f** since only three Ir layers are used, and the rearrangement of the Ir atom in the *top* layer exposes the gap

raise the temperature as the system is not connected to a bath and the energy cannot be dissipated. To control the temperature and remove the energy from the system a Langevin thermostat is used. The mechanism for this, as described in Sect. 2.8.2, involves the addition of both damping to the velocity and a driving force to a selection of atoms in the Ir surface (see below). In order to preserve the correct bulk dynamics for the system of interest a reasonable value for the damping parameter γ must be chosen.

7.3.1 Computation of Phonon DOS

In order to determine the damping parameter the dynamics of an iridium bulk lattice is simulated with a Langevin thermostat applied to a subset of the atoms positioned

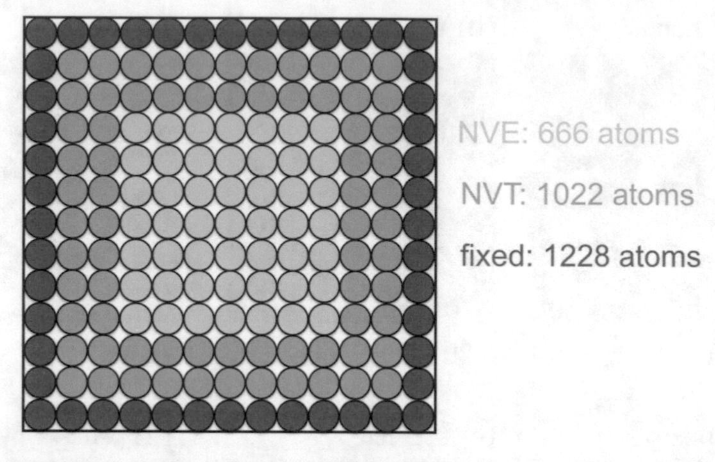

Fig. 7.3 A 2D illustration of the unit cell for the system for calculating the velocity autocorrelation function. The NVE atoms are surrounded by the NVT atoms with the Langevin thermostat applied. The outer layer consists of atoms fixed to the Ir bulk geometry

at the periphery of the simulation cell. From the dynamics the phonon DOS can be determined by taking the Fourier transform of the velocity autocorrelation function (VACF). Calculation of this is described in detail in Sect. 2.8.2.1. By varying the damping parameter the effect on the phonon DOS can be determined. This is compared with the phonon DOS as calculated by lattice dynamics (LD) in order to ensure that the choice of γ is suitable as explained below.

The three-dimensional unit cell is constructed as a system of layers surrounding a cube of NVE atoms. Around this there is a layer of NVT atoms with the Langevin thermostat applied. The outer layer of atoms is fixed to the Ir bulk lattice constant. This is required to ensure that the lattice potential of the internal atoms is correct. The unit cell, depicted in Fig. 7.3 consists of a total of 2916 Ir atoms. The cell must be large since many NVE atoms are required to calculate the VACF with good statistics. For the Ir interatomic potential the embedded atom model (EAM) is used [5] and the dynamics of the atoms themselves are simulated using the Large-Scale Atomic/Molecular Massively Parallel Simulator (LAMMPS) software [6]. The force field is used to determine the friction constant as the EAM model enables one to obtain the lattice dynamics of the Ir crystal reliably.

The NVT atoms are set to a temperature of 300 K. The time step is 1 fs. Initially the system is run for 1000 ps to equilibrate the temperature. After this the atomic velocities are sampled every 10 fs over a period of 100 ps. The VACF with a length of 20.48 ps is determined from this data. By taking the fast Fourier transform (FFT) of the VACF the phonon DOS is determined. This process is repeated for multiple values of the damping parameter. The VACF and the phonon DOS for $\gamma = 300\ \text{ps}^{-1}$, $\gamma = 100\ \text{ps}^{-1}$, $\gamma = 50\ \text{ps}^{-1}$, $\gamma = 10\ \text{ps}^{-1}$ are shown in Fig. 7.4a and b respectively.

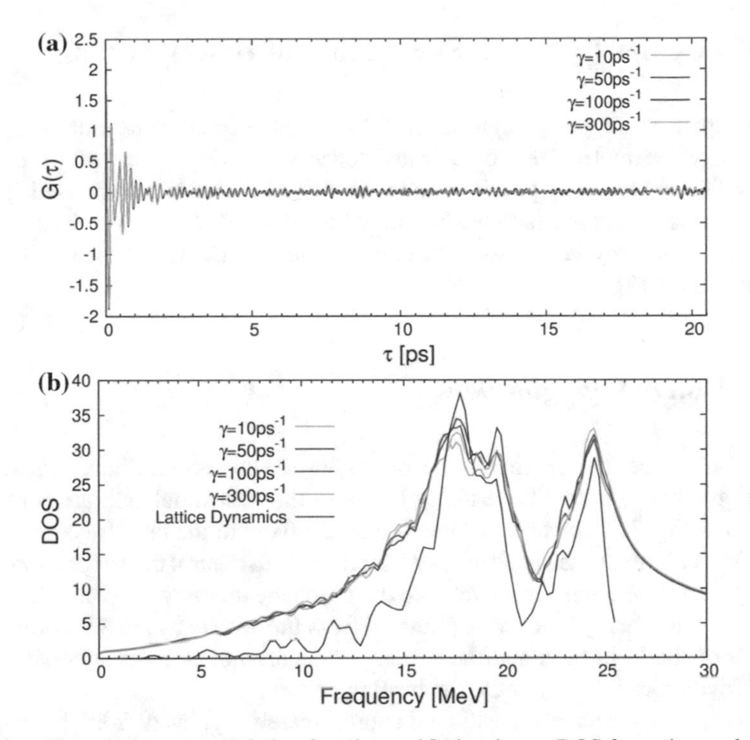

Fig. 7.4 **a** The velocity autocorrelation function and **b** the phonon DOS for various values of γ, and as determined from lattice dynamics

7.3.2 Choice of the Damping Parameter

In Fig. 7.4b the phonon DOS with different values of the damping parameter are compared to the phonon DOS as determined from zero temperature lattice dynamics calculations. These are performed with the General Utility Lattice Program (GULP) [7, 8] and the Ir EAM potential as before [5]. The phonon DOS, as determined from both lattice dynamics and MD simulations contains two main peaks. One of these is broad with its peak centred on 17 MeV. The other is narrower and centred on 24–25 MeV. The positions of these peaks matches the reference LD DOS regardless of the value of γ. The main difference between the LD DOS compared to the MD DOS is that the peaks are much narrower when determined from MD simulations. This is independent of the damping parameter. Based on the shape of the peak at 17 MeV, a damping parameter of 100 ps^{-1} gives the closest representation to the LD DOS. The peaks at 24 MeV for different γ values look similar to each other. However at $\gamma = 50$ and 100 ps^{-1} at small feature at 23 MeV is present, similar to the LD DOS. Therefore a damping parameter of 100 ps^{-1} is chosen for the Ir Langevin thermostat.

7.4 Molecular Dynamics Simulations of Defect Healing

The attempted healing of a single vacancy by a deposited ethylene molecule is simulated. Ethylene molecules are projected towards the defect site with a constant velocity. Different starting positions and velocities of the molecule are used to determine the conditions for which healing may be achieved. The Langevin thermostat in the graphene/Ir system is set to a temperature of 1123 K to match the growth conditions as in [2].

7.4.1 System Configuration

The SV Gr/Ir system, shown in Fig. 7.1a, consists of 3 layers of iridium, with a monolayer of graphene on top. The iridium layers in the MD simulation are configured as shown in Fig. 7.5. The bottom layer atoms are fixed to the bulk Ir geometry. The middle layer atoms are assigned the NVT thermostat to control the temperature of the entire system. The top layer atoms, like the graphene atoms, are described as NVE atoms. Thermal energy is added or removed from the system by the NVT atoms. The top layer of the Ir surface, and the carbon atoms are not thermostatted since their realistic dynamics is important to the healing process.

The ethylene molecule is positioned approximately 5 Å away from the graphene layer at the start of the simulation. It is initialised with atomic velocities drawn from the Maxwell distribution to match the system temperature of around 1123 K. In addition, the atomic velocities have a directional component towards the surface.

7.4.2 Initialisation

Prior to the deposition of the molecule towards the surface the surface must be initialised with the correct temperature and equilibrated over time. Initially, at $t = 0$ all velocities are set to zero. The thermostat is then applied to all surface atoms (except for the fixed Ir atoms) to set the temperature to 1123 K. The simulation is run until the temperature is stable. The temperature variation over time is shown in Fig. 7.6.

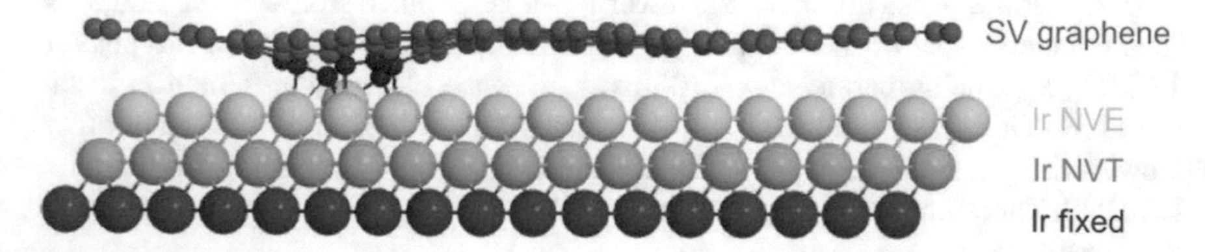

Fig. 7.5 The side view of the SV Gr/Ir system with 3 layers of iridium, the *middle* of which is attached to the NVT thermostat, while the atoms in the *bottom* layer are frozen

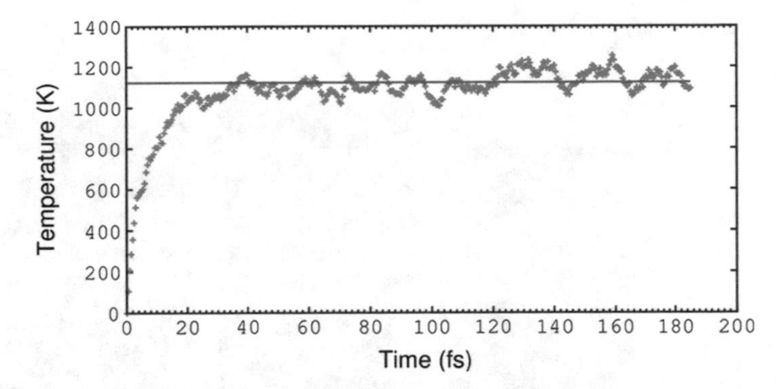

Fig. 7.6 The thermal equilibration of the SV Gr/Ir system. The surface system is initialised ensuring the temperature is set to 1123 K (*blue line*). Once equilibration has been reached the atomic velocities and positions are used to restart the MD simulations for the defect healing

After equilibration the final atomic positions and velocities are inserted as the initial conditions for the MD healing simulation. The ethylene molecule is initialised with Maxwell distributed velocities at the required temperature. An additional velocity component towards the surface is also added, the kinetic energy of this is discussed below. The initial state of the molecule is not equilibrated, however by the time it reaches the surface it is assumed that it will be. Its position and orientation with respect to the surface are chosen differently in each simulation.

7.4.3 Ethylene Molecule Deposition

The velocity of the ethylene molecule is given an initial value corresponding to an energy of 10–20 eV. This value is too large to be experimentally physical, however it is required to make the MD simulations of healing accessible in practice. The process of a molecule adsorbing onto graphene defect site requires the molecule to arrive at a very specific position otherwise the barrier for adsorption is large. In reality many molecules are projected towards the surface in a short time span and so such events may be possible even with a much lower incident velocity. In order to replicate this with MD simulations many different simulations with slightly different initial conditions would need to be run. This is impractical given the size of the system and the computational resources required for each MD run. Hence the molecule requires a large velocity to ensure it adsorbs to the surface.

7.4.4 Ethylene Molecule Starting Position

The ethylene molecules are positioned approximately 5 Å away from the top of the graphene layer. The molecule begins orientated in one of three different ways with

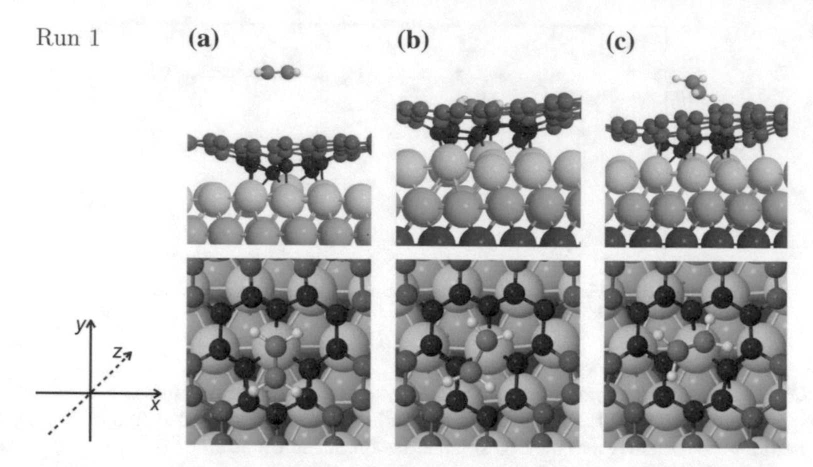

Fig. 7.7 Defect Run 1. The ethylene kinetic energy is 10 eV, and the molecule is centred at (0, 0) with respect to the centre of the defect site, and orientated along the y axis

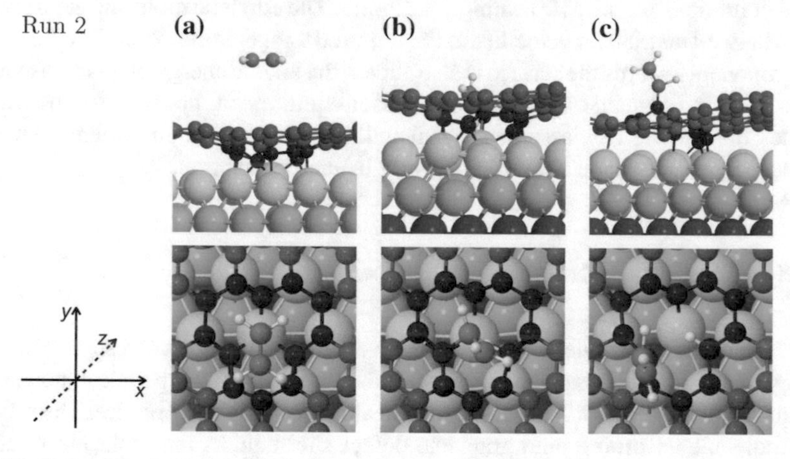

Fig. 7.8 Defect Run 2. The ethylene kinetic energy is 15 eV, and the molecule is centred at (0, 0) with respect to the centre of the defect site, and orientated along the y axis

respect to the surface plane: either aligned along the x, y, or z axes. The starting position (with respect to the defect site) is varied to alter where on the surface the molecule will collide with it. The reactions between the molecule and the defect site will depend on this as well as on the velocity, which will effect the healing process.

Figures 7.7, 7.8, 7.9, 7.10, 7.11, 7.12, 7.13, 7.14, 7.15 and 7.16 show the MD simulations of ethylene deposition towards a defect site. For each simulation the top and side views are shown for the start of the simulation (a), the point where the ethylene molecule reaches the surface at its closest point (b), and the end of

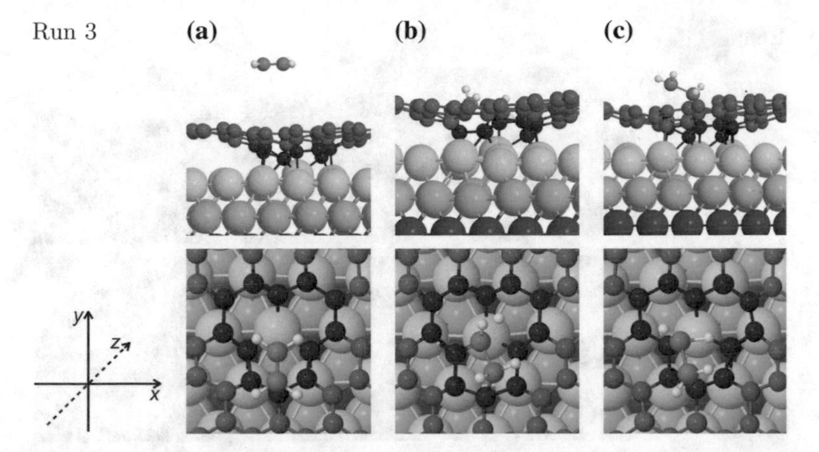

Fig. 7.9 Defect Run 3. The ethylene kinetic energy is 15 eV, and the molecule is centred at (0, –1Å) with respect to the centre of the defect site, and orientated along the y axis

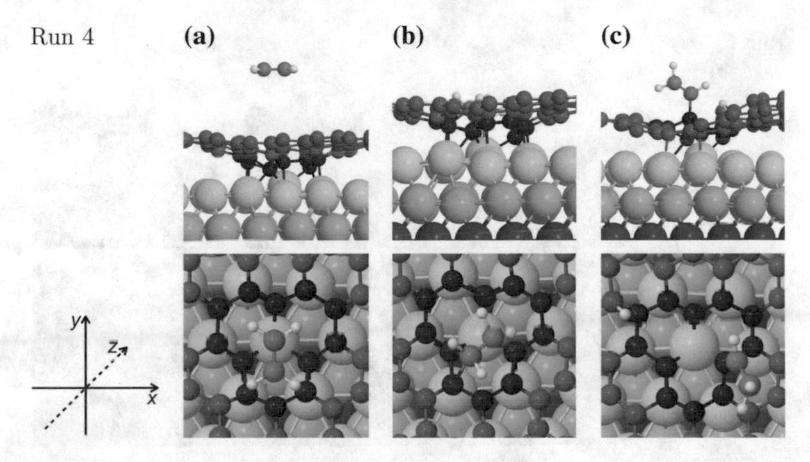

Fig. 7.10 Defect Run 4. The ethylene kinetic energy is 20 eV, and the molecule is centred at (0, 0) with respect to the centre of the defect site, and orientated along the y axis

the simulation (c) when either the molecule detaches from the surface, or remains attached with the thermal energy of the impact equilibrated throughout the system. Each simulation has a different set of initial conditions. The molecules' kinetic energy towards the surface, and its starting position and orientation are all varied. In Figs. 7.7, 7.8, 7.9, 7.10, 7.11, 7.12, 7.13, 7.14, 7.15 and 7.16 these are listed in terms of its position with respect to being directly above the defect site, and its orientation in terms of its C-C bond being aligned along the cartesian directions.

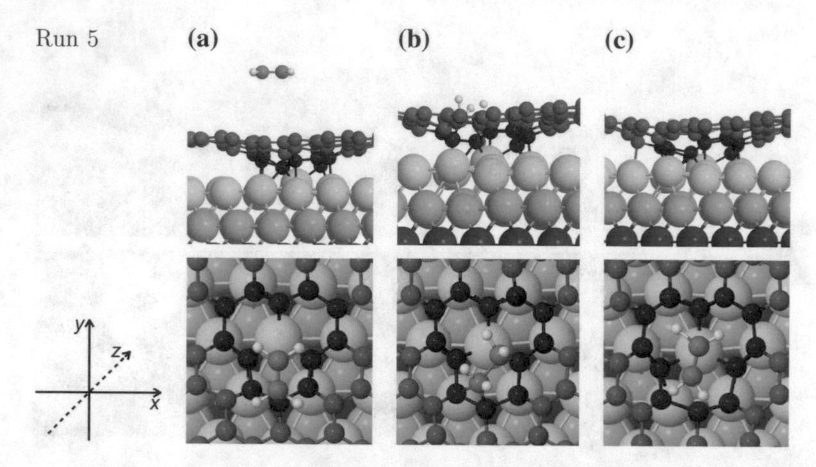

Fig. 7.11 Defect Run 5. The ethylene kinetic energy is 20 eV, and the molecule is centred at (0, –1Å) with respect to the centre of the defect site, and orientated along the y axis

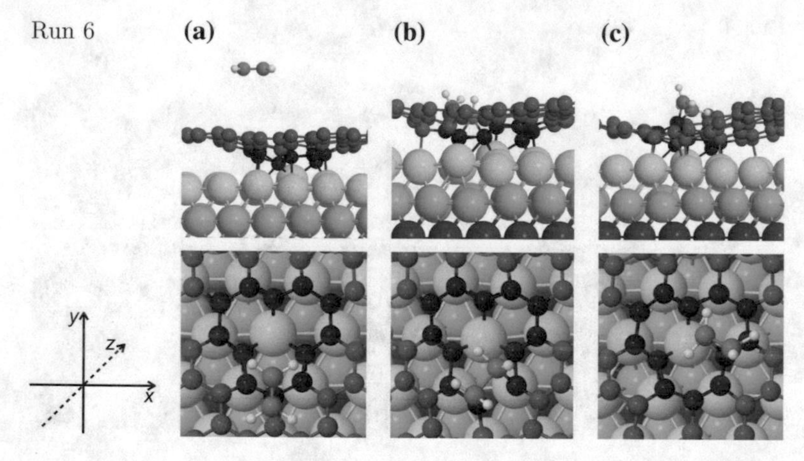

Fig. 7.12 Defect Run 6. The ethylene kinetic energy is 20 eV, and the molecule is centred at (0, –2Å) with respect to the centre of the defect site, and orientated along the y axis

7.4.5 Simulation Results

7.4.5.1 Run 1

The simulation begins with the ethylene molecule positioned directly above the defect site, orientated along the y-axis, as shown in Fig. 7.7a. It is projected towards the Gr/Ir surface with a kinetic energy of 10 eV. In Fig. 7.7b the molecule reaches the defect site. The hydrogen atoms are repelled away from the surface. The molecule is unable to bond to the surface and as a result returns backwards, moving slowly

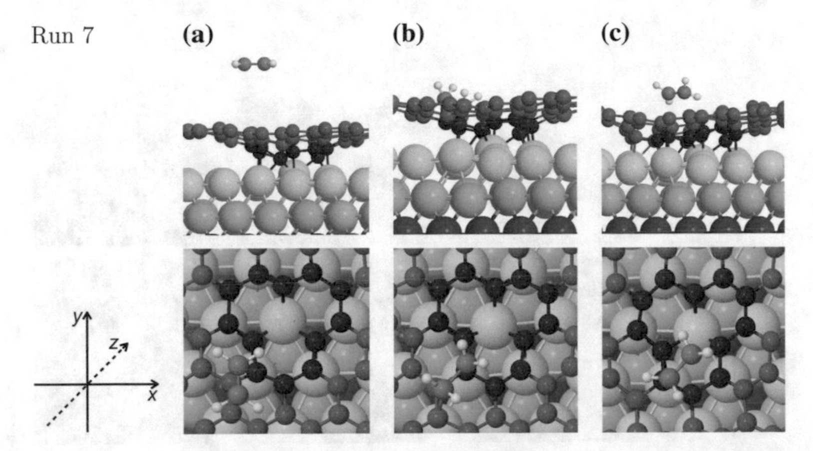

Fig. 7.13 Defect Run 7. The ethylene kinetic energy is 20 eV, and the molecule is centred at (–2Å, –2Å) with respect to the centre of the defect site, and orientated along the y axis

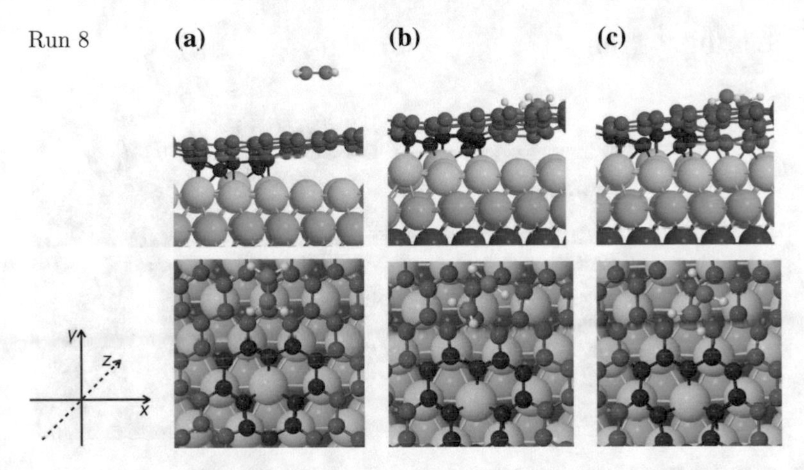

Fig. 7.14 Defect Run 8. The ethylene kinetic energy is 20 eV, and the molecule is centred at (0, +5Å) with respect to the centre of the defect site, and orientated along the y axis

away from the surface. The simulation is stopped when the molecule is a reasonable distance away, shown in Fig. 7.7c.

7.4.5.2 Run 2

The initial conditions of the simulation are the same as in Run 1, however the kinetic energy of the molecule's velocity component towards the Gr/Ir surface is increased to 15 eV. As it reaches the defect site it becomes more orientated along the z-axis. An H atom from the C atom on the righthand side of Fig. 7.8b is removed from the

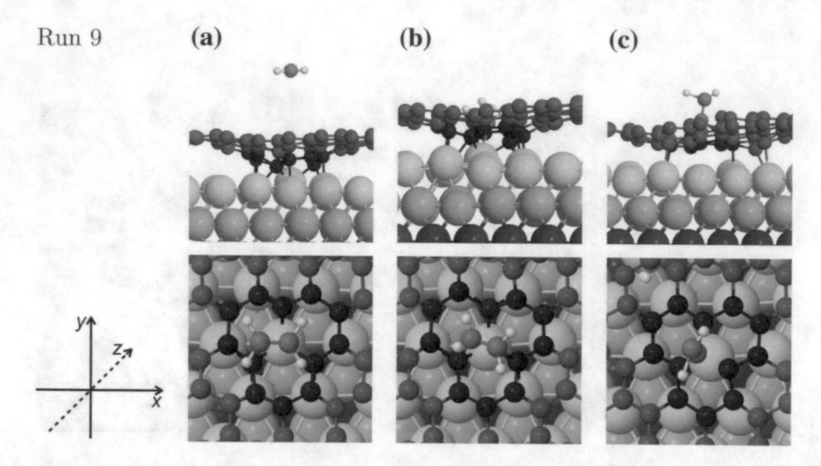

Fig. 7.15 Defect Run 9. The ethylene kinetic energy is 20 eV, and the molecule is centred at (0, 0) with respect to the centre of the defect site, and orientated along the x axis

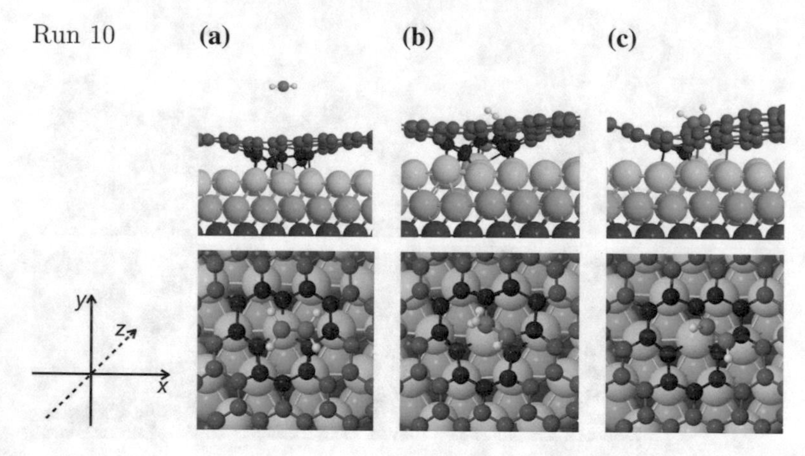

Fig. 7.16 Defect Run 10. The ethylene kinetic energy is 20 eV, and the molecule is centred at (+1Å, 0) with respect to the centre of the defect site, and orientated along the x axis

molecule to the nearby C atom, which surrounds the defect. In the absence of the H atom the molecule becomes attached to a C atom next to the defect, seen in Fig. 7.8c. Once the molecule's kinetic energy has been transferred to the surface atoms it begins to move upwards. It remains attached to the surface by a C defect atom, which is pulled upward with it.

7.4.5.3 Run 3

The simulation begins with the ethylene molecule offset from the defect site by -1Å in the y-direction as shown in Fig. 7.9a. The kinetic energy and orientation are the same as in Run 2. The molecule reaches the defect site and briefly attaches across the defect site, Fig. 7.9b. It then detaches from the surface and moves away from the surface, shown in Fig. 7.9c.

7.4.5.4 Run 4

The simulation set up is the same as in Run 1 and 2, as seen in Fig. 7.10a, except that the translational part of the kinetic energy of the molecule is increased to 20 eV. Figure 7.10b shows the molecule once it has reached the defect site. It attaches in the position shown and a H atom is removed from the right C atom to the graphene surface. It then diffuses to the position shown in Fig. 7.10c. The remaining ethylene structure eventually detaches at its left side, shown in Fig. 7.10c staying attached to the surface at its right side where the H was removed from. It positions itself away from the surface. This sequence of events is similar to that occurring in Run 2 where the initial position of the ethylene molecule was the same.

7.4.5.5 Run 5

The ethylene molecule begins offset from the defect site by -1Å in the y-direction, as in Run 3. However the translational velocity is increased to kinetic energy of 20 eV. The simulation events are also similar to those in Run 3. It reaches the surface as shown in Fig. 7.11b, but does not absorb in any way, and none of its bonds are broken. At the end of the simulation in Fig. 7.11c it has moved back away from the surface.

7.4.5.6 Run 6

The start of the simulation, shown in Fig. 7.12a begins with the ethylene molecule positioned -2 Å in the y-direction) away from being directly above the defect site, and with a kinetic energy of 20 eV. Upon reaching the surface the molecule is attracted towards the defect site as seen in Fig. 7.12b This shows that there may be some preference for the molecule to be at the defect site, rather than on the defect-free graphene. The molecule absorbs on top of a C atom, forming two C-C bonds as shown in Fig. 7.12c.

7.4.5.7 Run 7

Run 7 is similar to Run 6 except the distance between the molecules starting position and the point directly above the defect site is increased. At the start of the simulation the ethylene molecule begins offset from the defect site by 2 Å in both the x and y directions (Fig. 7.13a). During the simulation the molecule directs itself towards the defect site, moving slightly towards it. It briefly reaches the surface as shown in Fig. 7.13b and then begins to move back upwards (Fig. 7.13c). This is a similar result to Run 6.

7.4.5.8 Run 8

In this simulation the molecule is aimed away from the defect site in order to determine the range of the attractive forces between it and the defect site, which are demonstrated in Runs 6 and 7. During the simulation the molecule rotates slightly in order for the C atoms to align with C atoms in the graphene layer. The molecule does not adsorb or show any movement towards the defect site since it is too far away.

7.4.5.9 Run 9

In the previous simulations the ethylene molecule was orientated along the y-axis. In this simulation it is instead orientated along the x-axis as shown in Fig. 7.15a. Starting directly above the defect site the molecule is projected towards the defect with a kinetic energy of 20 eV. It reaches the surface maintaining a similar orientation and adsorbs as shown in Fig. 7.15b. Once the molecule has adsorbed two H atoms are removed from the C atom on the left to form an H_2(g) molecule. This moves upward into the gas phase. As the H atoms are removed the remaining C atom forms bonds between two of the defect-surrounding atoms. It then tilts upwards, detaching at the right side, to become vertical. The simulation ends with the molecule positioned vertically in the defect site as shown in Fig. 7.15c.

7.4.5.10 Run 10

This simulation has the same initial conditions as in the previous run, however the ethylene molecule begins shifted by 1Å to the right (Fig. 7.16a. The first half of the simulation is similar to Run 9, the molecule adsorbs across the defect site, and two H atoms are lost to form H_2(g). However instead one H atom is removed from each of the C atoms as shown in Fig. 7.16b. The remaining C atoms of the ethylene molecule form a hexagon with four of the C atoms that surround the defect site. In Fig. 7.16c one of the other C atoms is pushed down below the molecule and attaches to the Ir surface.

7.4.6 Conclusions from the MD Simulations

Based on the MD simulations some conclusions can be made about the initial stage of direct defect healing by ethylene deposition. The ethylene molecule is most likely to adsorb onto the defect site and remain there with stability if one or more H atoms are removed from it. In all but one cases where no C-H bonds where broken the ethylene molecule quickly desorbs from the surface after its initial kinetic energy component is transferred to the surface. In Runs 2, 4, 9 and 10 when the molecule is directed more or less directly at the defect site the molecule adsorbs.

In Runs 9 and 10, where the molecule is orientated along the x-axis two H atoms are removed from the molecule. This suggests that if the ethylene molecule arrives in such a way that it attaches across the defect site, bonding with two of the three innermost C atoms that neighbour the defect site, then the adsorption will be more likely to be successful. Runs 9 and 10 also end with the C atoms of the ethylene molecule positioned close to the defect, potentially in a position to fill the defect site.

If the molecule lands away from the defect site then it may be attracted towards the defect, as shown in Runs 6 and 7. In Run 6 the molecule is attracted to the one of the C atoms in the defect site where it absorbs without the removal of H atoms.

7.4.7 *Final States*

The MD runs where the ethylene molecule successfully absorbs onto the defect site possibly represent intermediate states towards the healing of the defect. The MD simulations are run for an adequate amount of time to ensure that the system is stable. The geometry of the final state is then relaxed. The relaxed geometries of the systems resulting from MD runs 2, 4, 6, 9 and 10 are shown in Fig. 7.17a–e, respectively.

For each of the systems in Fig. 7.17, the formation energy is determined in terms of the energy difference between the SV Gr/Ir system with the ethylene molecule

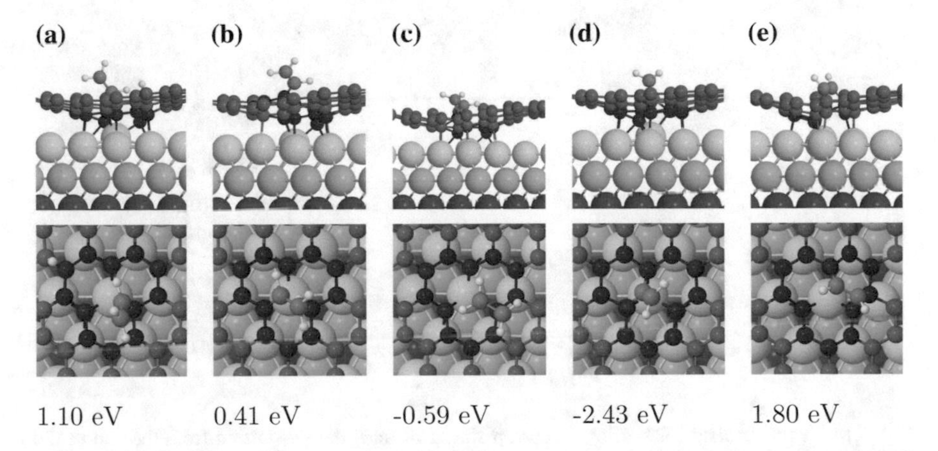

Fig. 7.17 The relaxed geometries for partially healed Gr/Ir defect systems as determined from MD simulations: **a** Run 2, **b** Run 4, **c** Run 6, **d** Run 9 and **e** Run 10

adsorbed $E_{SV+C_2H_4}$, and the SV Gr/Ir, E_{SV} system plus the ethylene molecule in the gas phase, $E_{C_2H_4(g)}$:

$$E_{form} = E_{SV+C_2H_4} - E_{SV} - E_{C_2H_4(g)}. \tag{7.1}$$

For Runs 9 and 10 $E_{SV+C_2H_4}$ is replaced by $E_{SV+C_2H_2} + E_{H_2(g)}$ since two of the H atoms form H_2(g). The formation energies are also shown in Fig. 7.17.

The configuration in Fig. 7.17d, where H_2 is formed and the remains of the ethylene molecule attaches in the defect site vertically is by far the most energetically stable. It is also the geometry closest to healing the defect, since the lower C atom completes one of the three hexagons that surround the defect. The healing pathway of this configuration is investigated in Sect. 7.5, below. The other configurations are less energetically favourable. This may be because none of the hexagonal units are completed with the addition of the molecule.

7.5 NEB Healing of the Single Vacancy Defect

Starting from the geometry in Fig. 7.17d a mechanism to produce a healed graphene structure is sought. The simplest way to achieve this is to incorporate the bottom C atom (originally from the ethylene molecule) into the graphene structure to form three complete carbon hexagons. The remaining atoms from the ethylene molecule (CH_2) must either desorb from the graphene or remain on the graphene in a stable configuration. Since CH_2 is unstable in the gas phase it is likely that it will not easily desorb in this case. However in an actual experiment there will likely be other atoms or defects nearby, allowing the possibility for the CH_2 to either react with two H atoms to form methane gas, or diffuse and be incorporated into another defect. For

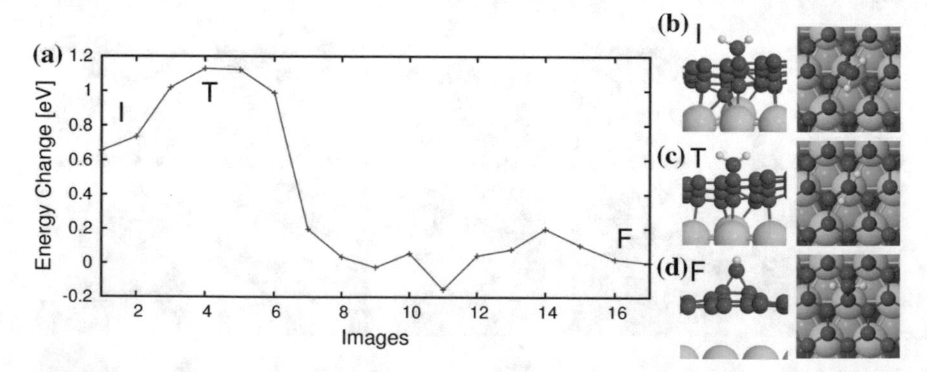

Fig. 7.18 A mechanism for healing a hcp-top single vacancy defect starting from the end of the MD simulation in Fig. 7.17d. **a** The energy profile, **b** the initial state **c** the transition state and **d** the final state are shown. The energy barrier for the healing is 0.47 eV and the reverse barrier is 1.13 eV

the purpose of this study the CH_2 in the final state is placed on top of the healed graphene sheet, above a C-C bond where it is most stable. The final (healed) state is shown in Fig. 7.18d.

The NEB calculated energy profile for healing of the defect is shown in Fig. 7.18a. Note that the final state is 0.66 eV lower in energy than the initial state but it is still not the global minimum since the presence of the CH_2 molecule of top of the graphene increases the energy of the final state. To fully complete the healed graphene this would need to be removed by one of the processes described above. During the healing process the C atoms surrounding the defect site move upwards away from the iridium surface and the single vacancy is filled by the C atom from ethylene. As this happens the graphene detaches from the surface and the energy of the system is lowered. The energy barrier for this process is 0.47 eV, which is small given that the experimental temperature is around 1120 K. This means that this healing process should be possible. However, there are more processes involved in this healing mechanism, such as the energy barrier for the adsorption of the ethylene molecule itself, which is a complicated process involving the formation of an H_2 molecule. The mechanism for this will be investigated with NEB calculations in the future.

7.6 Conclusions

In this chapter the healing single vacancy defect in graphene grown on the Ir(111) surface was investigated using a combination of molecular dynamics simulations and NEB calculations. In particular the healing of the hcp-top single vacancy was studied since this was found to have the lowest formation energy out of the six single vacancy defects that may form in the high symmetry regions on the graphene/Ir(111) moiré unit cell.

MD simulations of deposition of ethylene molecules, with their motion directed towards the defect site were performed. In these simulations the Langevin thermostat was used in order to control the temperature and remove energy to the system. The value of the damping parameter was chosen by computing the phonon DOS for bulk iridium using different values for the damping parameter, and then selecting the one which most closely resembles the phonon DOS as determined from lattice dynamics.

Multiple MD healing simulations were run with different starting parameters for the ethylene molecule. Of the ten performed five ended with the ethylene molecule absorbed on the defect site, and in some cases H_2(g) was lost from the molecule upon its collision with the graphene. From these five successful runs the lowest energy (relaxed) structure at the end of the run was selected as a starting point to investigate the full healing mechanism in detail using NEB. The NEB results found a 0.47 eV barrier for part of the healing process, which should be easily overcome at the experimental temperatures. However, there are further stages involved in this healing mechanism which should be investigated in the future and different healing mechanisms should also be studied.

References

1. Q. Yu, L.A. Jauregui, W. Wu, R. Colby, J. Tian, Z. Su, H. Cao, Z. Liu, D. Pandey, D. Wei et al., Control and characterization of individual grains and grain boundaries in graphene grown by chemical vapour deposition. Nat. Mater. **10**(6), 443–449 (2011)
2. N. Blanc, F. Jean, A.V. Krasheninnikov, G. Renaud, J. Coraux, Strains induced by point defects in graphene on a metal. Phys. Rev. Lett. **111**, 085501 (2013)
3. F. Banhart, J. Kotakoski, A.V. Krasheninnikov, Structural defects in graphene. ACS Nano **5**(1), 26–41 (2011)
4. S. Standop, O. Lehtinen, C. Herbig, G. Lewes-Malandrakis, F. Craes, J. Kotakoski, T. Michely, A.V. Krasheninnikov, Carsten Busse, Ion impacts on graphene/ir(111): Interface channeling, vacancy funnels, and a nanomesh. Nano Lett. **13**(5), 1948–1955 (2013)
5. H.W. Sheng, M.J. Kramer, A. Cadien, T. Fujita, M.W. Chen, Highly optimized embedded-atom-method potentials for fourteen fcc metals. Phys. Rev. B **83**, 134118 (2011)
6. S. Plimpton, Fast parallel algorithms for short-range molecular dynamics. J. Comput. Phys. **117**, 1–19 (1995)
7. J.D. Gale, GULP - a computer program for the symmetry adapted simulation of solids. JCS Faraday Trans. **93**, 629 (1997)
8. J.D. Gale, A.L. Rohl, The general utility lattice program. Molecul. Simul. **29**, 291–341 (2003)

Chapter 8
Final Remarks

8.1 Conclusions

This thesis presented a theoretical study about the growth of epitaxial graphene on the Ir(111) surface. The main area of focus was the early stages of graphene growth, where typically hydrocarbon species thermally decompose on the surface to form the C-species that are required for forming graphene. Following from this carbon clusters can nucleate on the surface, which then go on to form graphene islands. Specifically the thermal decomposition of ethylene on the Ir(111) surface and the nucleation of carbon clusters were studied in detail. Healing of graphene defects when graphene is grown on the Ir(111) surface was also investigated. In order to study these processes a variety of theoretical techniques were employed. Most of these were based on density functional theory (DFT) calculations. These methods were discussed in detail along with specific details pertaining to the Ir(111) system that was used in the calculations.

The subject of the first main chapter was the thermal decomposition mechanism for of ethylene on the Ir(111) surface under a temperature ramp as in the early stages of temperature programmed graphene growth. This was investigated with a combined experimental and theoretical approach. The theoretical method began with the formulation of a reaction scheme incorporating all possible reaction processes starting from ethylene and including all $C_{1-2}H_{1-6}$ hydrocarbon species. In order to understand the most preferable mechanism from this scheme the rates of each reaction were required. These were determined by finding the relevant energy barriers with the nudged elastic band method and then using the Arrhenius equation. For the NEB calculations the initial and final reaction states were found by finding the lowest energy configuration of each hydrocarbon species. The reaction rates were then used in rate equations in order to determine the kinetics of the decomposition and find the evolution of species on the surface.

In-situ XPS experiments were also used to study the thermal decomposition. In order to help identify the peaks in the XPS spectra core level binding energy

H.A. Tetlow, *Theoretical Modeling of Epitaxial Graphene Growth on the Ir(111) Surface*, Springer Theses, DOI 10.1007/978-3-319-65972-5_8

calculations of the C 1s level of each species were performed. From this the coverage of species on the surface over the course of a temperature ramp was found.

Based on both the experimental results and the rate equations results the following decomposition mechanism was found:

$CH_2CH_2 \rightarrow CH_2CH + H$
$CH_2CH \longleftrightarrow CH_2C + H$
$CH_2C + H \longleftrightarrow CH_3C$
$CH_2CH \rightarrow CHCH + H$
$CHCH \rightarrow CH + CH$
$CH \rightarrow C + H$

The limiting process in the formation of C monomers is the formation of H_2(g), which needs to occur in order for H atoms to be removed from the surface so that CH can dehydrogenate to C. Breaking of the C-C bond is an important step since it allows for the formation of C monomers rather than dimers. This means that carbon clusters may then be formed containing an odd number of C atoms. This result supports the experimental work in [1] where carbon clusters containing five C atoms were suggested to be heavily involved in graphene growth.

Although both the experimental and theoretical results give the same overall reaction mechanism for the ethylene decomposition it was found that there were differences in the relative concentrations of some of the species, especially CH_3C and CHCH. This was attributed to the fact that rate equations do not contain any spatial information.

In order to include spatial effects into the kinetics a lattice based kinetic Monte Carlo code for the thermal decomposition of ethylene was devised. This was the subject of Chap. 4. In addition to the reaction processes already considered it was also necessary to understand the diffusion of species in order to incorporate spatial effects into the kMC simulation. The diffusion barriers of each species were determined with NEB calculations. The diffusion reactions can be incorporated as proper kMC moves, however it was found that since diffusion moves are on average high rate processes they happen much faster, which slows down the progress of the simulation considerably. To speed up the simulation it was demonstrated that diffusion could be incorporated by randomly moving the species on the surface to the same effect.

In Chap. 5 the kMC code was used to simulate the thermal decomposition of hydrocarbons under different conditions, in addition to the decomposition of ethylene decomposition with a temperature ramp that was already discussed. These included using different fixed temperatures as well as constant dosing of one species throughout the process (as in CVD), and the use of different species and substrates (such as methane and Pt(111)). Both high and low coverages of species were considered in each case.

By comparing the decomposition of ethylene as determined by the kMC simulation with the experimental results it was found that there were still differences in the relative coverages of species over the course of the temperature ramp, even at low coverages when any interactions between species are minimal. These discrepancies were suggested to be due to small errors in the DFT energy barriers. Changing some of these values slightly was found to give a much better agreement between the experimental results and the kMC simulations. The kMC simulation results were also compared with those of the rate equations, using the same parameters. At low coverages the rate equations and kMC gave similar results, however this was not the case at high coverages. This is due to the lack of spatial information in rate equations, which will naturally become more crucial at higher coverages.

The kMC simulations were performed considering Pt(111) as the substrate instead of Ir(111). Energy barriers for the reactions were taken from [2]. The decomposition mechanism was found to be the same as for the Ir(111) surface, however the relative coverages of species over the same temperature ranges were different. This is due to difference in the amount of H atoms on the surface. H atoms are lost from the Pt(111) at a lower temperature than on Ir(111) and hence hydrogenation reactions will be limited. Hence, it was found that due to the lack of H atoms C monomers are produced more readily than on Ir(111).

Finally the kMC code was then used to simulate the thermal decomposition of hydrocarbons that are constantly dosed onto surface as in CVD. For ethylene there was little change in the results from the TPG results. When methane was deposited it was found that it must dehydrogenate immediately to CH_3. As with ethylene, the formation of C monomers from methane was limited by the presence of H atoms, which allow C monomers to rehydrogenate easily to CH.

Following from the formation of C monomers the next step in graphene growth is the nucleation of carbon clusters on the surface. This was discussed in Chap. 6. The novel part of this study was the derivation of the temperature dependent cluster work of formation. This can be used to determine the critical cluster size N^* for which the addition of C atoms is equally favourable compared to their removal, and the corresponding nucleation barrier. This includes terms dependent on the cluster's energy, as well as its vibrational free energy and rotational degrees of freedom.

The work of formation was calculated for a variety of different carbon clusters, with size ranging from $N = 1 - 16$. These were split into five different structural types: chains, arches, rings, top-hollow compact or dome-like clusters. The results for the work of formation over a range of temperatures demonstrated the importance of temperature effects. The critical cluster size and nucleation barrier were found to vary over the temperature ranges studied. Up to 490 K it was determined that there are two critical cluster sizes at $N^* = 4$ and $N^* = 10$, suggesting that clusters with sizes between $N = 5 - 6$ should be stable on the surface, and may be able to contribute directly to the graphene growth front, as suggested in [1]. Furthermore as the temperature increases the nucleation barrier associated with $N = 10$ increases, making growth by monomer attachment unfavourable at high temperatures. This indicates that cluster growth may be more likely to occur by the coalescence of small stable clusters.

From observation of the change in the work of formation over the range of N studied it is noticeable that different types of cluster may be stable over difference sizes. For small N ($N < 10$) linear clusters such as chains and arches are most stable, but for larger N compact clusters (especially dome-like clusters) become more stable. Structural reconstructions between different cluster types were investigated with NEB calculations. The results suggested that this may be possible for some cluster types, however the transitions may have large barriers if many bonds are required to break during the reconstruction. Forming such clusters may instead be possible by coalescence of the long lived C_5 and C_6 clusters.

In Chap. 7 the healing of single vacancy defects in graphene on the Ir(111) surface was investigated using a combination of MD simulations and NEB calculations. Of six possible defect structures on the graphene/Ir(111) moiré lattice the top-hcp defect was selected since it was found to have the lowest formation energy.

MD simulations were performed in an attempt to heal the defect. The decomposition of an ethylene molecule towards the defect site in experimental conditions was simulated. This required the use of a Langevin thermostat for which a suitable value for the damping parameter was determined. Ten different MD runs were performed, each with different starting conditions. From these the structure with the lowest energy (after relaxation) at the end of the run was selected as a starting point for investigating the full healing mechanism using NEB. From these calculations it was found that the healing of this defect required overcoming a 0.47 eV energy barrier, which should be easily overcome given the experimental temperature. However there are more processes that are required to fully heal the defect that should be investigated in future as well as other healing mechanisms.

8.2 Limitations and Further Work

Overall the results and conclusions achieved in this thesis come from computational methods where approximations are made.

In all the DFT calculations performed lateral interactions between neighbouring species are excluded, since such a treatment would be impractical considering the almost infinite number of possible ways for the different species to interact. In the kinetic simulations these interactions will affect the values of the energy barriers in a way that cannot be easily predicted, and hence the results of the kMC simulations and the rate equations will differ to the experimental results.

In Chap. 3 the energy barriers for the reactions involved in the decomposition of ethylene on the Ir(111) surface were calculated using the DFT based NEB method. The accuracy of these barriers is crucial for correctly determining the reaction rates and hence simulating the reaction kinetics with both rate equations and kMC. However as previously discussed in Chap. 5 the results for the thermal evolution of the species does not completely match with the experimental results, even at low coverages where lateral interactions should be minimal. This suggested that the values of the energy barriers were subject to small errors. Therefore it was necessary to adjust these barriers slightly to get a good agreement with the experimental results. Although only small adjustments were made, knowing in which way to adjust the barriers would not have been possible without the experimental results, and therefore our results would be less accurate without the experimental comparison.

Another source of error in the reaction rates comes from the pre-exponential factors. In most cases these were taken to be $10^{13}s^{-1}$ however for the most crucial reaction these were calculated. To improve the accuracy the pre-exponential factors for each reaction could be calculated, and with a reasonable number of layers. However this would be too computationally expensive.

All the DFT calculations are limited by the number of Ir layers used. In general four layers have been used for most of the calculations in this thesis. However, as shown in the Appendix Sect. A.1.3, the number of layers can affect the energy barriers and pre-exponential factors somewhat. If possible it would have been desirable to use many layers for these calculations, however this would become too computationally expensive. In general improving the accuracy of the DFT calculations has a trade off between accuracy and computational effort.

The kMC simulations are subject to approximations as well. As discussed in Chap. 4 the diffusion of species is treated by randomly moving each species at each time step. The test simulations performed in Sect. 4.6 showed that this approximation is reasonable, however it would be more accurate to include the diffusion with proper kMC treatment. This was not possible since it slowed down the simulation too much. The kMC simulation also use a lattice model, whereby each species is constrained to a particular lattice site (or selection of sites) on the lattice. This assumes that species exist only in their lowest energy configuration. In reality this may differ, especially at high coverages when a particular site may be fully occupied.

Despite all of these limitations the rate equations and kMC simulation for the ethylene decomposition on the Ir(111) surface still manage to predict the correct reaction sequence as compared with the experimental results. Furthermore when compared to the rate equations the kMC results improve on the agreement showing that the kMC simulations give a good description of the coverage effects that are not included within the rate equations.

In the future this kMC code could be easily applied for investigating the decomposition of hydrocarbons on other fcc (111) substrates, as was done with Pt(111) in Sect. 5.6. As a second stage the growth of carbon clusters from monomers could be simulated by including further reactions into the reaction scheme. This, however would be complicated by the many reaction possibilities as the clusters continue to grow in size. In would also be of interest to perform this study with a completely different hydrocarbon, such as benzene, in order to determine whether C monomers are also produced during the decomposition.

The nucleation study in Chap. 6 could be improved by considering more cluster types as this would make it a more complete study. However around over 50 different clusters were included and many of the stable clusters found agree with previous theoretical studies on Ir(111) [3, 4]. As a future work larger clusters could be investigated as well as the coalescence of small clusters, which was suggested as a possible method of cluster formation in the current study.

The healing of defects in graphene was investigated for only one SV defect type. Further work could include more defect types. Ten MD healing runs were performed altogether; however if possible more should be considered in order to ensure the best healing mechanism is found. The MD simulations represent a simplified version of the experimental situation since there may be many more ethylene molecules and defects in one given region, which increases the number of possible processes that may occur during healing. Including all possibilities would be impractical. To further this study more NEB calculations are required to investigate additional healing processes.

References

1. E. Loginova, N.C. Bartelt, P.J. Feibelman, K.F. McCarty, Evidence for growth by C cluster attachment. New J. Phys. **10**(093026) (2008)
2. Y. Chen, D.G. Vlachos, Hydrogenation of ethylene and dehydrogenation and hydrogenolysis of ethane on Pt(111) and Pt(211): a density functional theory study. J. Phys. Chem. C **114**(11), 4973–4982 (2010)
3. P. Wu, H. Jiang, W. Zhang, Z. Li, Z. Hou, J. Yang, Lattice mismatch induced nonlinear growth of graphene. J. Am. Chem. Soc. **134**, 6045–6051 (2012)
4. C. Herbig, E.H. Åhlgren, W. Jolie, C. Busse, J. Kotakoski, A.V. Krasheninnikov, T. Michely, Interfacial carbon nanoplatelet formation by ion irradiation of graphene on iridium(111). ACS Nano **8**(12), 12208–12218 (2014)

Appendix A
Hydrocarbon Decomposition

A.1 Lowest Energy Hydrocarbon Geometries

To obtain the relaxed geometries each species is placed onto the Ir(111) surface slab. Four Ir layers (the bottom two of which fixed to the bulk geometry) are used.

For the NEB energy barrier calculations the lowest energy geometries for the hydrocarbon species are required. In order to ensure this different trial structures are considered, each optimised to find their geometry. The geometry with the lowest adsorption energy is then selected. An example of this is shown in Fig. A.1 for CH_3CH.

After finding the lowest energy configurations for each species the adsorption energy is re-calculated with the BSSE correction [1].

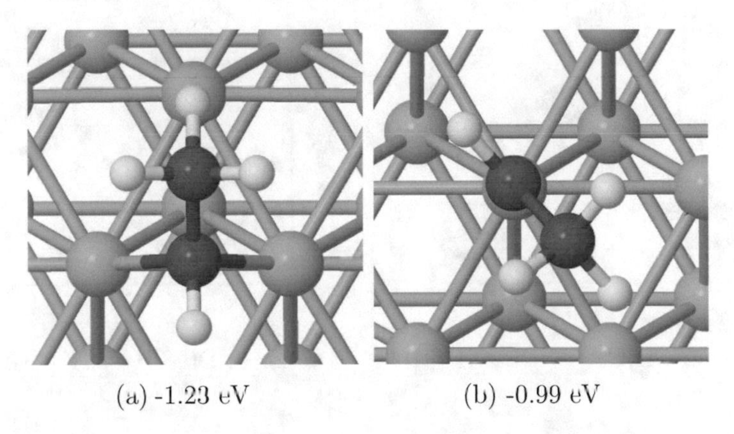

Fig. A.1 The two geometries found for CH_3CH. The binding energy for the geometry in **a** is 0.24 eV lower than in **b**. Hence the geometry in **a** is taken to be the default geometry for CH_3CH in all other calculations

H.A. Tetlow, *Theoretical Modeling of Epitaxial Graphene Growth on the Ir(111) Surface*, Springer Theses, DOI 10.1007/978-3-319-65972-5

A.1.1 NEB Reaction Profiles

The NEB reaction profiles for the reactions in Chap. 3 are shown below.

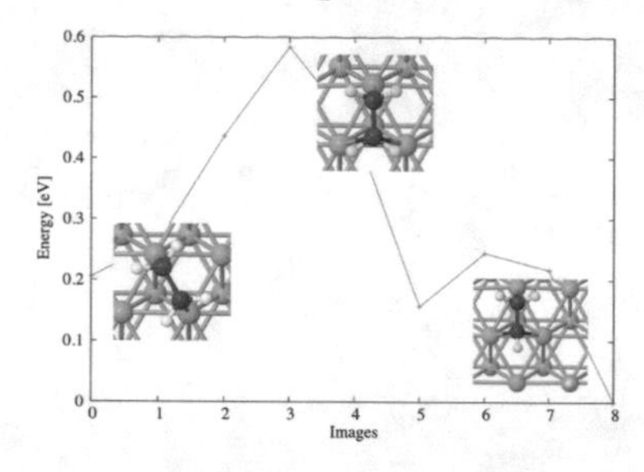

Reaction 1: $CH_2CH_2 \rightarrow CH_2CH + H$

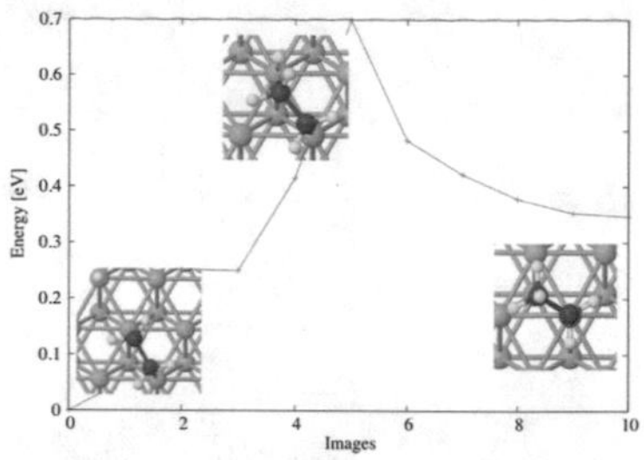

Reaction 2: $CH_2CH_2 + H \rightarrow CH_3CH_2$

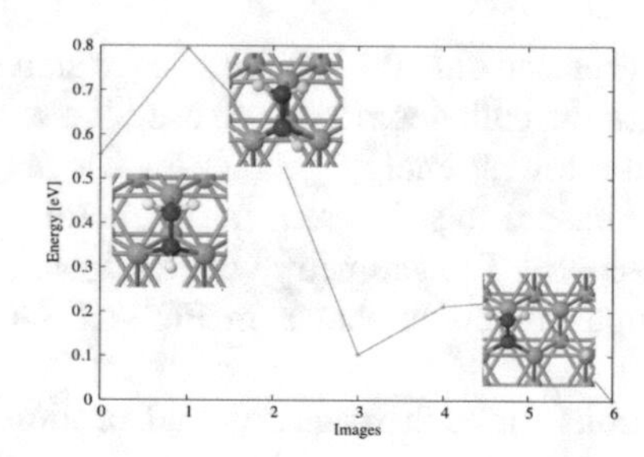

Reaction 3: $CH_2CH \rightarrow CH_2C + H$

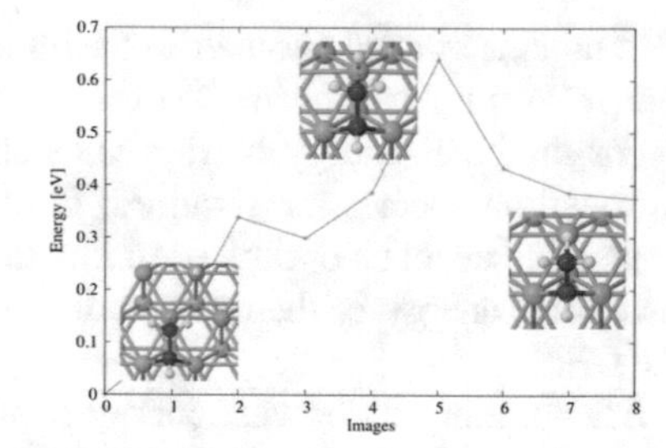

Reaction 4: $CH_2CH + H \rightarrow CH_3CH$

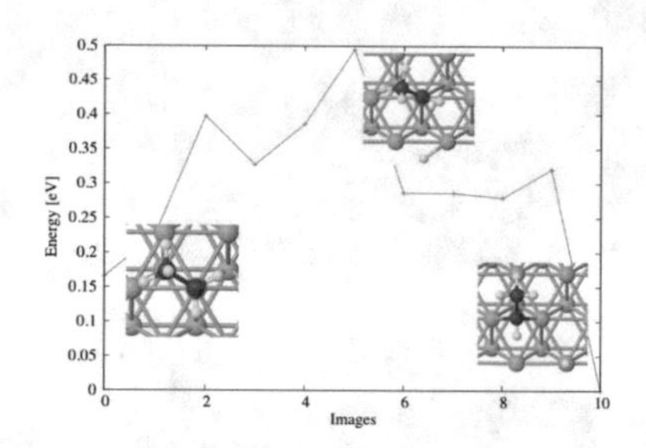

Reaction 5: $CH_3CH_2 \rightarrow CH_3CH + H$

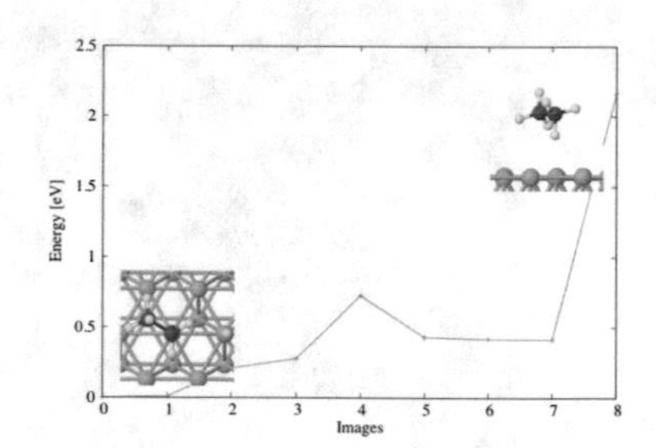

Reaction 6: $CH_3CH_2 + H \rightarrow CH_3CH_3$

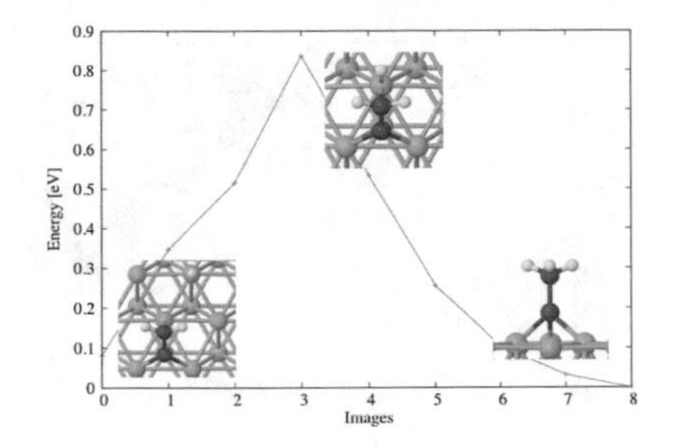

Reaction 7: $CH_2C + H \rightarrow CH_3C$

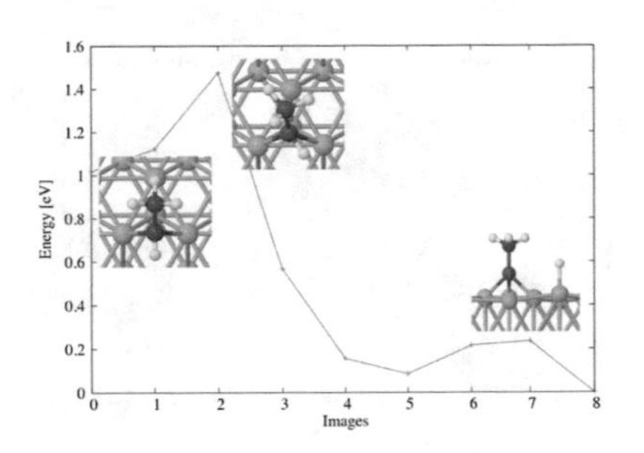

Reaction 8: $CH_3CH \rightarrow CH_3C + H$

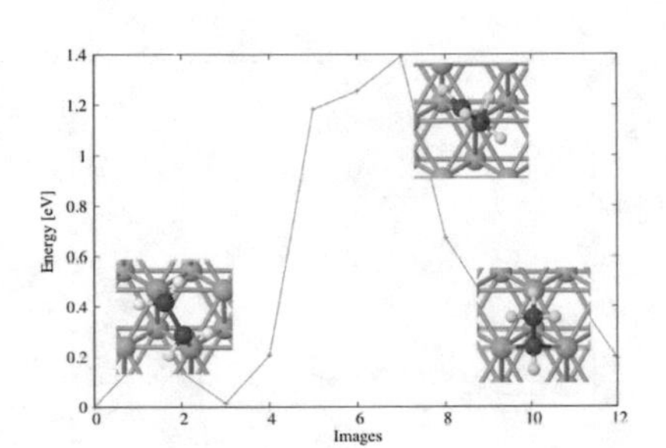

Reaction 9: $CH_2CH_2 \rightarrow CH_3CH$

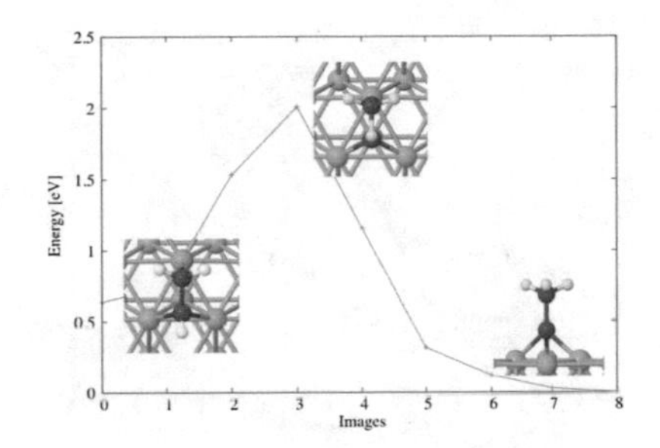

Reaction 10: $CH_2CH \rightarrow CH_3C$

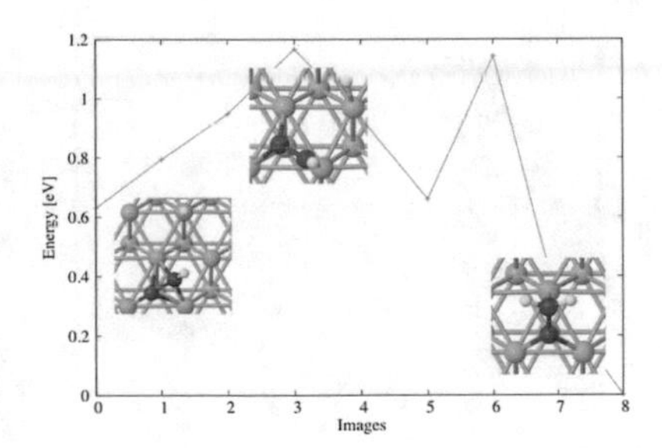

Reaction 11: $CHC + H \rightarrow CH_2C$

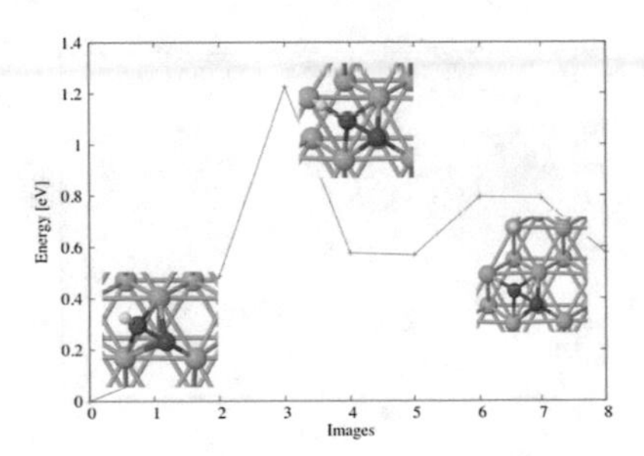

Reaction 12: $CHC \rightarrow CC + H$

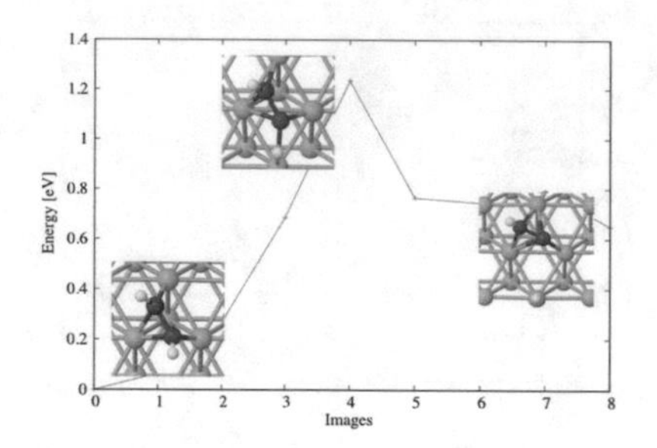

Reaction 13: CHCH $\rightarrow$ CHC + H

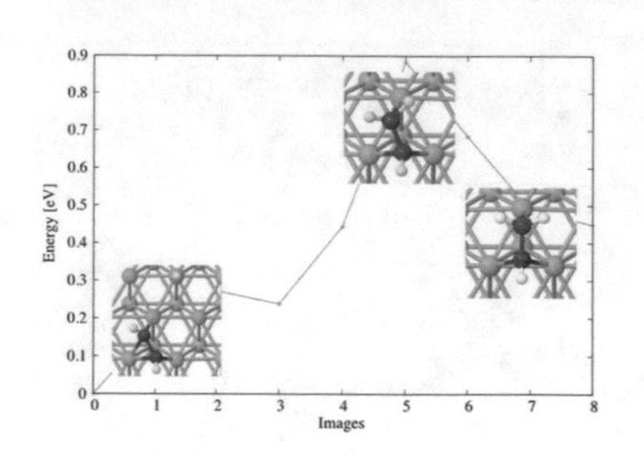

Reaction 14: CHCH +H $\rightarrow$ CH_2CH

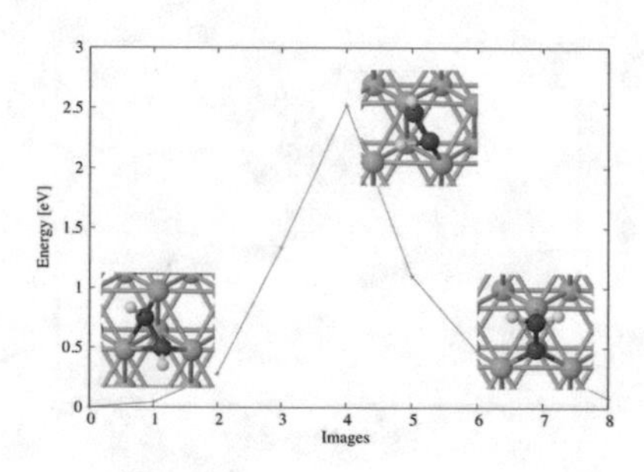

Reaction 15: CHCH $\rightarrow$ CH_2C

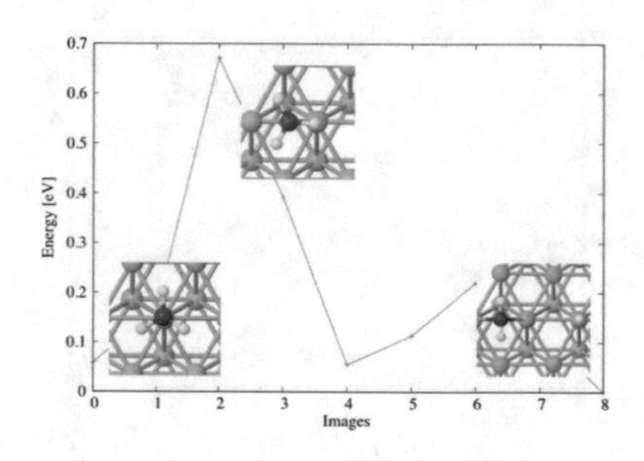

Reaction 16: CH_3 $\rightarrow$ CH_2 + H

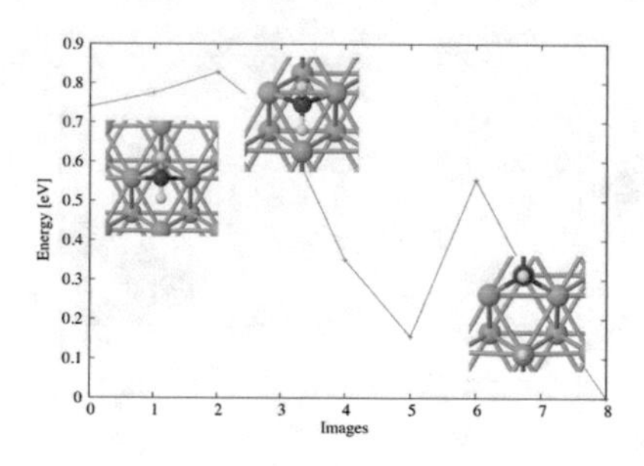

Reaction 17: CH_2 $\rightarrow$ CH + H

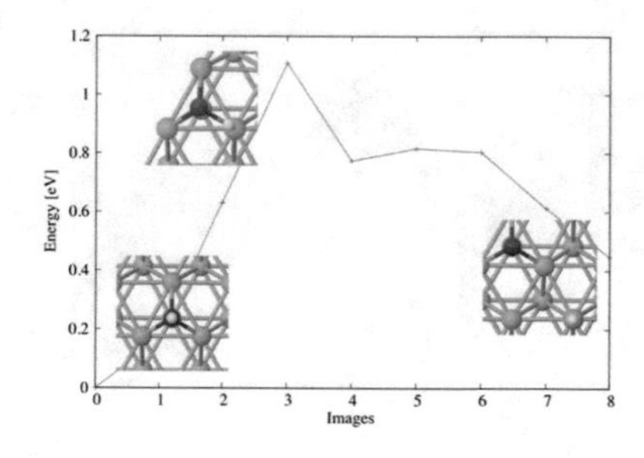

Reaction 18: CH $\rightarrow$ C + H

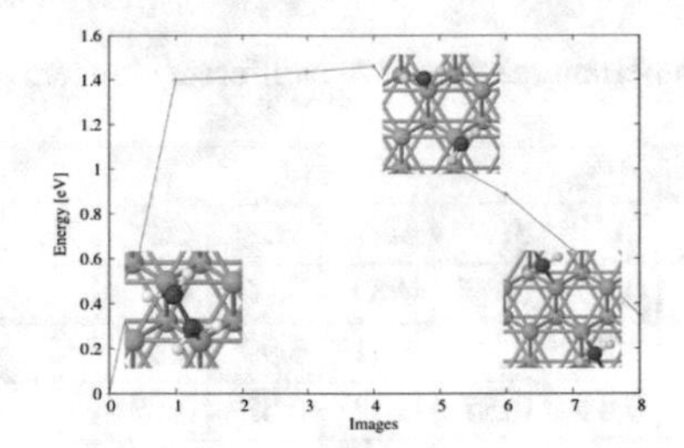

Reaction CB1: $CH_2CH_2 \rightarrow CH_2 + CH_2$

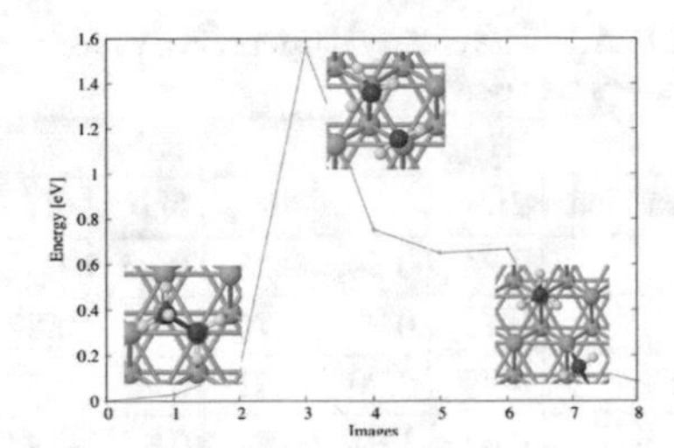

Reaction CB2: $CH_3CH_2 \rightarrow CH_3 + CH_2$

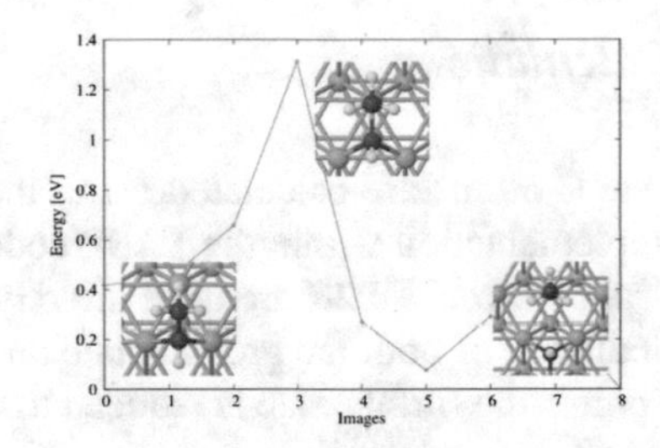

Reaction CB3: $CH_3CH \rightarrow CH_3 + CH$

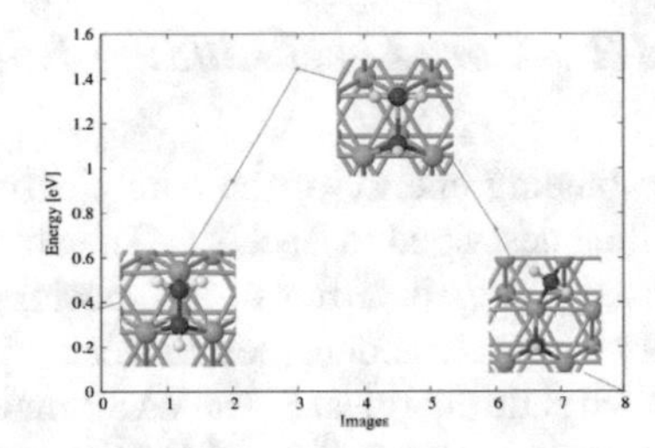

Reaction CB4: $CH_2CH \rightarrow CH_2 + CH$

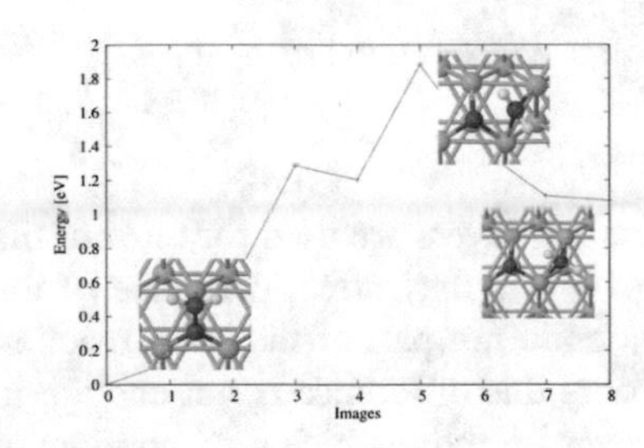

Reaction CB5: $CH_2C \rightarrow CH_2 + C$

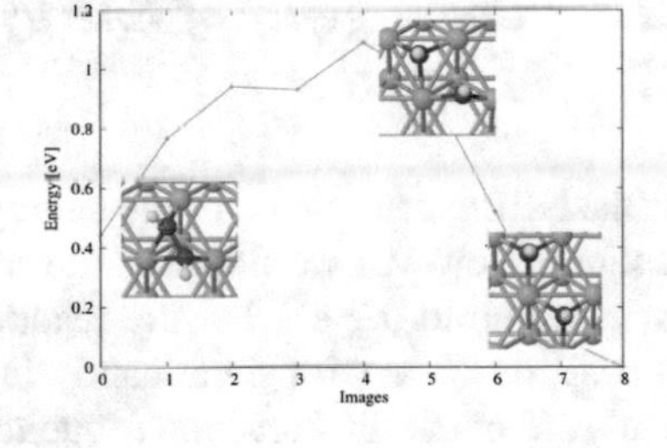

Reaction CB6: $CHCH \rightarrow CH + CH$

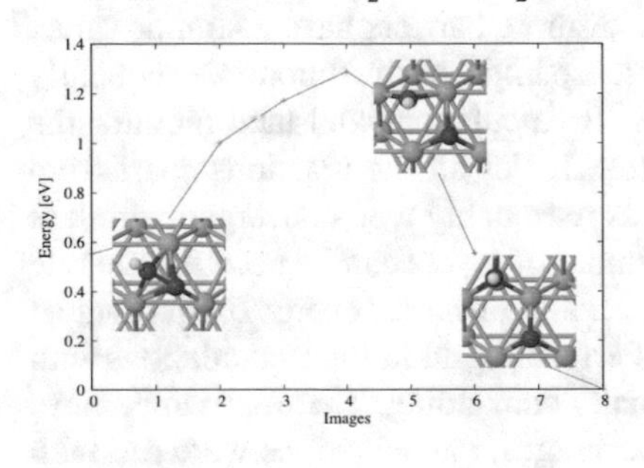

Reaction CB7: $CHC \rightarrow CH + C$

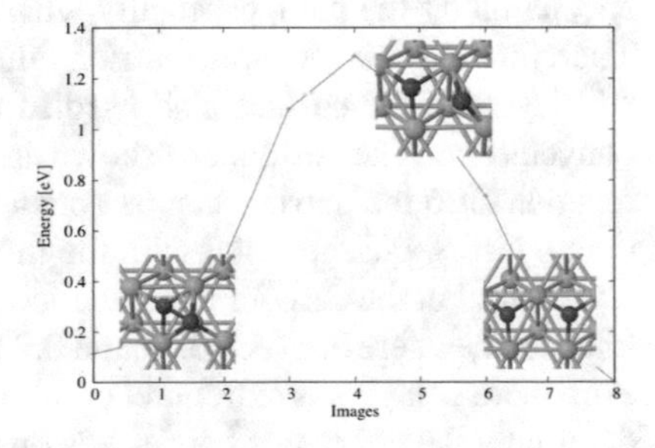

Reaction CB8: $CC \rightarrow C + C$

Table A.1 The energy barriers (in eV) for the most important reactions when different numbers of Ir layers are used

	E_{for}						E_{back}					
Reaction	2L	3L	4L	5L	6L	7L	2L	3L	4L	5L	6L	7L
3	0.24	0.34	0.38	0.35	–	–	0.79	0.65	0.59	0.66	–	–
7	0.76	0.91	0.76	0.84	0.82	–	0.84	0.99	1.01	0.99	0.99	–
14	0.88	0.80	0.71	0.80	0.77	–	0.44	0.55	0.52	0.53	0.53	–
CB6	0.65	1.04	0.82	0.87	0.93	0.79	1.09	1.22	1.15	1.15	1.15	1.13

A.1.2 Core Level Binding Energy Calculations

The binding energy of the core electrons in the C atoms are calculated using the method described in Sect. 2.6. The standard implementation within the VASP code is used. The geometries of the species as determined from CP2K are used directly. Only the electron densities are relaxed for calculation of both the ground state and excited state energies. Since VASP includes **k** points the surface slab is reduced to 4 $\times$ 4 atoms with a 2 $\times$ 2 $\times$ 1 k point sampling.

A.1.3 Convergence of Energy Barriers with Number of Layers

For the NEB calculations for the energy barriers 2 Ir layers are used for most of the reactions. However we find that the number of layers may affect the value of the barrier by up to 0.2 eV. For the reactions which are not part of the main reaction pathway, or those with significantly large barriers this difference is not enough to greatly affect the final reaction sequence. However for those which are important to the reaction pathway this energy difference can affect the coverages of the various species along the path, especially where two or more barriers have a similar value. Therefore to get an accurate barrier value for these important reactions we sequently add layers to the surface slab used in the NEB calculations and then monitor the convergence. The results are shown in Table A.1. For all the reactions apart from reaction CB6 the barriers can be considered as reasonably well converged with less than 6 layers and the result with the most number of layers can be used for the rate equations. For the CB6 forward reaction however the converged energy barrier is most likely somewhere between 0.79 and 0.93 eV. Performing additional calculations with even more Ir layers is extremely computationally demanding. Therefore, in order to determine the effect this barrier has on the coverages, rate equations were run with different values of the barrier within this energy range, and then compared with the experimental results. The results of these simulations are shown in Fig. A.2 for energy barriers of 0.79, 0.85 and 0.93 eV.

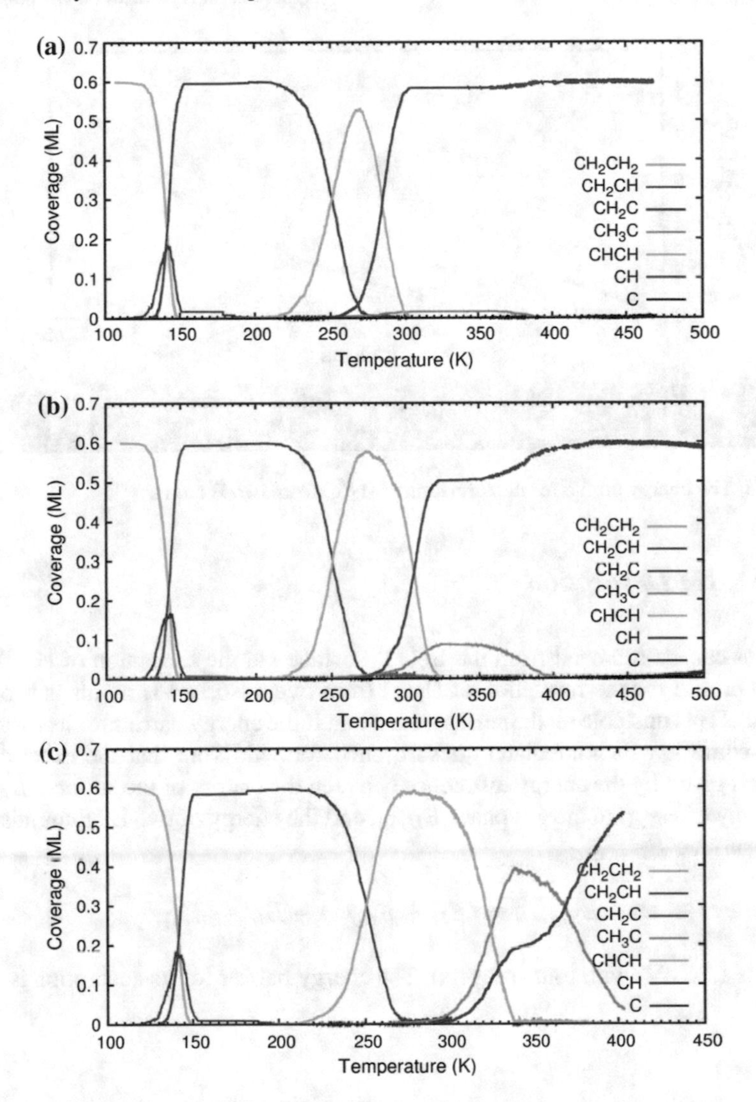

Fig. A.2 Temperature evolution of species coverage determined from kMC simulations where the energy barrier for the forward reaction of CB6 are **a** 0.79 eV, **b** 0.85 eV, **c** 0.93 eV

When increasing the barrier for the CB6 reaction (CHCH $\rightarrow$ CH + CH) from 0.79 to 0.93 eV we find that the lifetime of CHCH is increased by about 40 K. This is desirable since the experimental results predict a long lifetime for CHCH. However excessively increasing the barrier results in the formation of CH_3C. This is because it is possible to form CH_3C before CHCH can convert into CH. Based on this the energy barrier is taken as 0.85 eV.

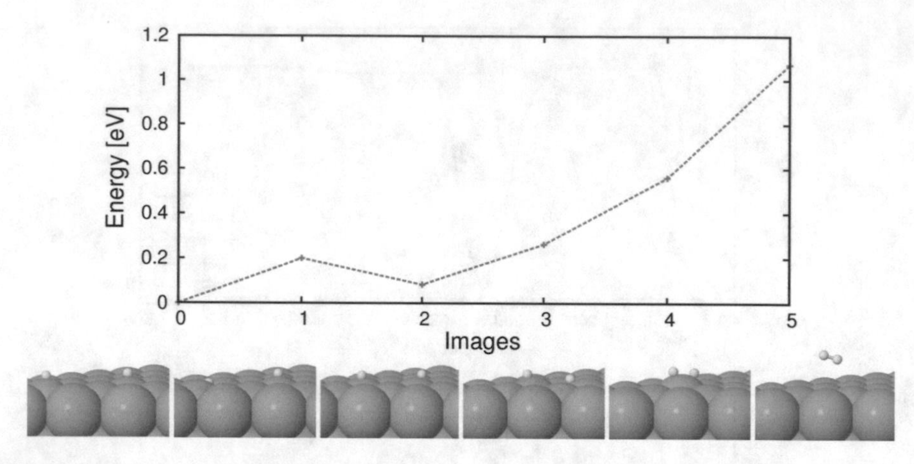

Fig. A.3 The energy profile for the formation of H_2(g) from two H atoms

A.1.4 H_2 Desorption

H atoms can be removed from the Ir(111) surface via the formation of H_2(g). The energy profile for the formation of H_2(g) from two adsorbed H atoms is shown in Fig. A.3. H_2 is unstable on the surface, such that if the energy barrier for its formation is overcome it will immediately desorb once formed. Note that the exact energy barrier is given by the energy difference between the energy of the surface, E_{Ir} plus the energy of H_2(g) in the gas phase $E_{H_2(g)}$, and the energy of two H atoms adsorbed on the surface $E_{Ir+2H(ads)}$:

$$E_{H_2 form} = (E_{Ir} + E_{H_2(g)}) - E_{Ir+2H(ads)} \tag{A.1}$$

which is 1.25 eV (with four Ir layers). The energy barrier for its adsorption is negligible.

A.1.5 Pre-exponential Factors

Calculating the reaction rates from the energy barriers involves the use of a pre-exponential factor ν, which represents the attempt frequency for the reaction. For this value it is reasonable to take 10^{13} s^{-1}, a value which has an order of magnitude of the atomic vibrational frequency. However for a few key reactions which are important for the reaction mechanism we determined the value of the pre-exponential factors in the equation for their rates. This was done by calculating the vibrational frequencies in the initial, final and transition states and using Vineyard's formula. This is explained in Sect. 2.7.1.

Table A.2 The calculated pre-exponential factors ν for 2 and 3 iridium layers

Reaction	$\nu_{i,j(for)}$ s^{-1} (2lay)	$\nu_{i,j(for)}$ s^{-1} (3lay)	$\nu_{j,i(back)}$ s^{-1} (2lay)	$\nu_{j,i(back)}$ s^{-1} (3lay)
3	1.7×10^{12}	–	5.3×10^{12}	–
7	1.7×10^{13}	8.9×10^{12}	1.2×10^{12}	1.9×10^{12}
14	2.6×10^{13}	1.6×10^{12}	7.8×10^{12}	3.1×10^{12}
CB6	3.6×10^{12}	–	5.9×10^{11}	–

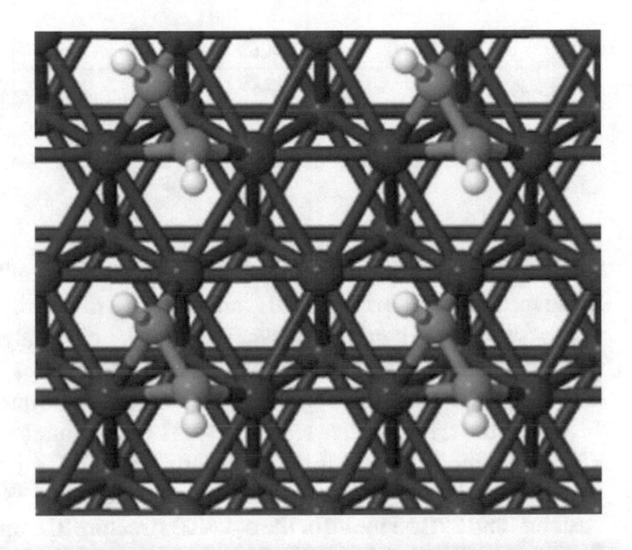

Fig. A.4 The c(4 × 2) superstructure as shown in the case of CHCH

As with the energy barriers we also check for convergence with the number of Ir layers for reactions 7 and 14. The pre-exponential factors are shown in Table A.2. The pre-exponential factors do not affect the rates as much as the energy barriers so even a change by one order of magnitude will not greatly change the coverage results.

A.1.6 Vibrational Frequency Calculations and Coverage Effects

From observation of Fig. 3.6 it can be seen that there are several peaks differing in their position by about 405 meV that can be interpreted as belonging to vibrational satellites associated with a single species. The LEED results also show a superstructure of adsorbed molecules, representing the c(4 × 2) structure for at this same temperature range. This arrangement is shown in Fig. A.4 for CHCH.

To analyse this superstructure and its vibrational satellites we calculated the core level shifts and the vibrational frequencies of the surface modes of the core excited

Table A.3 The calculated core level shifts and vibrational frequencies for CH-CH, CH and CH_3 with the c(4 × 2) superstructure on the Ir surface. For CH-CH the two frequencies correspond to the stretching of the C-H bonds with (1) the core shifted C atom and (2) the regular C atom. For the CH the single frequency is from the stretching of the C-H bond. CH_3 has three frequencies due to (1) the symmetric stretching of all three C-H bonds, (2) the asymmetric stretching of two of the C-H bonds, and (3) two of the C-H bonds stretching in-phase, while the remaining C-H bond stretches out-of-phase

Species	E ($n_c - 1$) [eV]	Vibrational frequencies [meV]
CH-CH	–54.57	(1) 343.0
		(2) 378.5
CH	–54.63	341.1
CH_3	–55.49	(1) 340.9
		(2) 343.7
		(3) 343.9

final state for species containing either a single C atom or two equivalent C atoms. Of these species we rule out CH_2CH_2 and CH_2 due to their instability in this temperature range (see energy barriers). Instead we restrict our analysis to the CH-CH, CH and CH_3 molecules. The calculated vibrational frequencies for different coverages are shown in Table A.3 for the species with a C core electron shifted to the valence band.

Out of the three species tested CH-CH has a vibrational frequency closest to the vibrational mode seen in the XPS spectra. For the C atom which is not core shifted the C-H bond gives rise to a vibrational frequency of 381.55 meV for the c(4 × 4) structure and 378.55 meV for the c(4 × 2) structure. Compared to the other calculated frequencies that are in the 340–350 meV range this is the most reasonable candidate for the species found experimentally.

A.1.7 Rate Equations

In our formulation for the rate equations we consider a system of fifteen ordinary differential equations for the concentration of each of the species, N_i in Table A.4 and the concentration of hydrogen, N_H. The terms in the equations for each species represent the reactions it is involved in. Each reaction will provide a positive (concentration increasing) and negative (concentration decreasing) term into the rate equation. Positive terms depend on the concentration of any species which are required to form the current species as well as the corresponding reaction rate. Negative terms depend on the concentration of the current species and any other species with which it reacts, and also the rate for that reaction. The rate equations for each species are listed below. Note that the rates for reactions where the C-C bond is broken are denoted $R_{i(j),k}$ where N_i and N_j are single C species CH_m.

Table A.4 The labels corresponding to hydrocarbon species as used in the rate equations

Index	Species
1	CH_2CH_2
2	CH_3CH_2
3	CH_2CH
4	CH_3CH_3
5	CH_3CH
6	CH_2C
7	CH_3C
8	CHC
9	CHCH
10	CC
11	CH_3
12	CH_2
13	CH
14	C

$$\frac{dN_1}{dt} = -R_{1,2}N_1N_H + R_{2,1}N_2 + R_{3,1}N_3N_H - R_{1,3}N_1 - R_{1,5}N_1 + R_{5,1}N_5 \\ +R_{12+12_1,1}N_{12}^2 - R_{1,12+12}N_1$$

$$\frac{dN_2}{dt} = R_{1,2}N_1N_H - R_{2,1}N_2 - R_{2,4}N_2N_H - R_{2,5}N_2 + R_{5,2}N_5N_H \\ +R_{11+12,2}N_{11}N_{12} - R_{2,11+12}N_2$$

$$\frac{dN_3}{dt} = R_{1,3}N_1 - R_{3,1}N_3N_H - R_{3,5}N_3N_H + R_{5,3}N_5 - R_{3,6}N_3 + R_{6,3}N_6N_H \\ -R_{3,7}N_3 + R_{7,3}N_7 + R_{12+13,3}N_{12}N_{13} - R_{3,12+13}N_3 - R_{3,9}N_3 + R_{9,3}N_9N_3$$

$$\frac{dN_4}{dt} = R_{2,4}N_2N_H$$

$$\frac{dN_5}{dt} = R_{1,5}N_1 - R_{5,1}N_5 + R_{2,5}N_2 - R_{5,2}N_5N_H + R_{3,5}N_3N_H - R_{5,3}N_5 - R_{5,7}N_5 \\ +R_{7,5}N_7N_H + R_{11+13,5}N_{11}N_{13} - R_{5,11+13}N_5$$

$$\frac{dN_6}{dt} = R_{3,6}N_3 - R_{6,3}N_6N_H + R_{6,7}N_6N_H - R_{7,6}N_7 - R_{6,9}N_6 + R_{9,6}N_9 - R_{6,8}N_6 \\ + R_{8,6}N_8N_H + R_{12+14,6}N_{12}N_{14} - R_{6,12+14}N_6$$

$$\frac{dN_7}{dt} = R_{3,7}N_3 - R_{7,3}N_7 + R_{5,7}N_5 - R_{7,5}N_7N_H + R_{7,6}N_7 - R_{6,7}N_6N_H$$

$$\frac{dN_8}{dt} = R_{6,8}N_6 - R_{8,6}N_8N_H - R_{8,9}N_HN_8 + R_{9,8}N_9 - R_{8,10}N_8 + R_{10,8}N_{10}N_H + R_{13+14,8}N_{14}N_{13} - R_{8,13+14}N_8$$

$$\frac{dN_9}{dt} = R_{3,9}N_3 - R_{9,3}N_HN_9 + R_{6,9}N_6 - R_{9,6}N_9 + R_{8,9}N_8N_H - R_{9,8}N_9 + R_{13+13,9}N_{13}^2 - R_{9,13+13}N_9$$

$$\frac{dN_{10}}{dt} = R_{8,10}N_8 - R_{10,8}N_{10}N_H + R_{14+14,10}N_{14}^2 - R_{10,14+14}N_{10}$$

$$\frac{dN_{11}}{dt} = -R_{11+12,2}N_{11}N_{12} + R_{2,11+12}N_2 - R_{11+13,5}N_{11}N_{13} + R_{5,11+13}N_5 + R_{12,11}N_{12}N_H - R_{11,12}N_{11}$$

$$\frac{dN_{12}}{dt} = -R_{12+13,3}N_{12}N_{13} + R_{3,12+13}N_3 - R_{12+11,2}N_{11}N_{12} + R_{2,12+11}N_2 - 2R_{12+12,1}N_{12}^2 + 2R_{1,12+12}N_1 - R_{12+14,6}N_{12}N_{14} + R_{6,12+14}N_6 - R_{12,11}N_{12}N_H + R_{11,12}N_{11} + R_{13,12}N_{13}N_H - R_{12,13}N_{12}$$

$$\frac{dN_{13}}{dt} = -R_{13+12,3}N_{13}N_{12} + R_{3,13+12}N_3 - R_{13+11,5}N_{11}N_{13} + R_{5,13+11}N_5 - R_{13+14,8}N_{13}N_{14} + R_{8,13+14}N_8 - 2R_{13+13,9}N_{13}^2 + 2R_{9,13+13}N_9 - R_{13,12}N_{13}N_H + R_{12,13}N_{12} + R_{14,13}N_{14}N_H - R_{13,14}N_{13}$$

$$\frac{dN_{14}}{dt} = -R_{14+12,6}N_{14}N_{12} + R_{6,14+12}N_6 - R_{14+13,8}N_{13}N_{14} + R_{8,14+13}N_8 - 2R_{14+14,10}N_{14}^2 + 2R_{10,14+14}N_{10} - R_{14,13}N_{14}N_H + R_{13,14}N_{13}$$

$$\frac{dN_H}{dt} = -R_{1,2}N_1N_H + R_{2,1}N_2 - R_{3,1}N_3N_H + R_{1,3}N_1 - R_{2,4}N_2N_H + R_{2,5}N_2 - R_{5,2}N_5N_H + R_{3,5}N_3N_H - R_{5,3}N_5 + R_{3,6}N_3 - R_{6,3}N_6N_H + R_{5,7}N_5 - R_{7,5}N_7N_H - R_{6,7}N_6N_H + R_{7,6}N_7 + R_{6,8}N_6 - R_{8,6}N_8N_H + R_{3,9}N_3 - R_{9,3}N_9N_H - R_{8,9}N_HN_8 + R_{9,8}N_9 + R_{8,10}N_8 - R_{10,8}N_{10}N_H - R_{12,11}N_{12}N_H + R_{11,12}N_{11} - R_{13,12}N_{13}N_H + R_{12,13}N_{12} - R_{14,13}N_{14}N_H + R_{14,13}N_{13} - 2R_HN_H^2$$

Appendix B
Carbon Clusters and Their Formation Energy at $T = 0$

All the carbon clusters considered in Chap. 6 and the their formation energies (at $T = 0$) are shown below, in order of ascending size N.

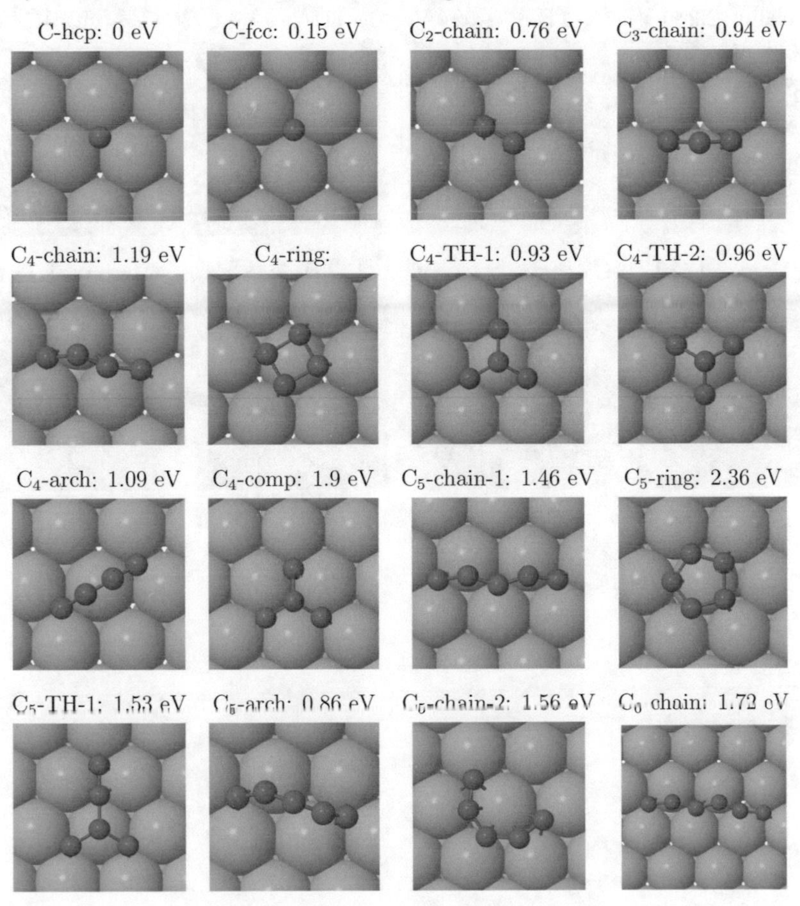

H.A. Tetlow, *Theoretical Modeling of Epitaxial Graphene Growth on the Ir(111) Surface*, Springer Theses, DOI 10.1007/978-3-319-65972-5

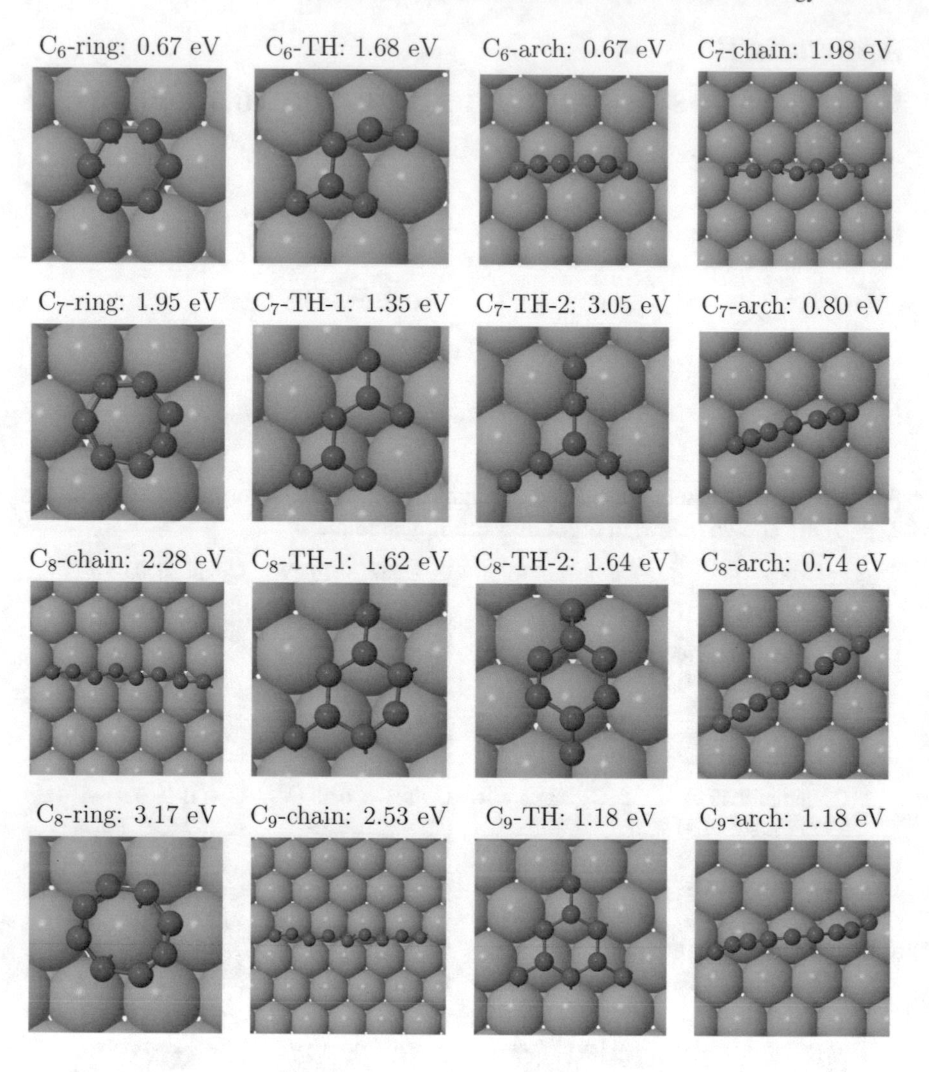
C_6-ring: 0.67 eV
C_6-TH: 1.68 eV
C_6-arch: 0.67 eV
C_7-chain: 1.98 eV
C_7-ring: 1.95 eV
C_7-TH-1: 1.35 eV
C_7-TH-2: 3.05 eV
C_7-arch: 0.80 eV
C_8-chain: 2.28 eV
C_8-TH-1: 1.62 eV
C_8-TH-2: 1.64 eV
C_8-arch: 0.74 eV
C_8-ring: 3.17 eV
C_9-chain: 2.53 eV
C_9-TH: 1.18 eV
C_9-arch: 1.18 eV

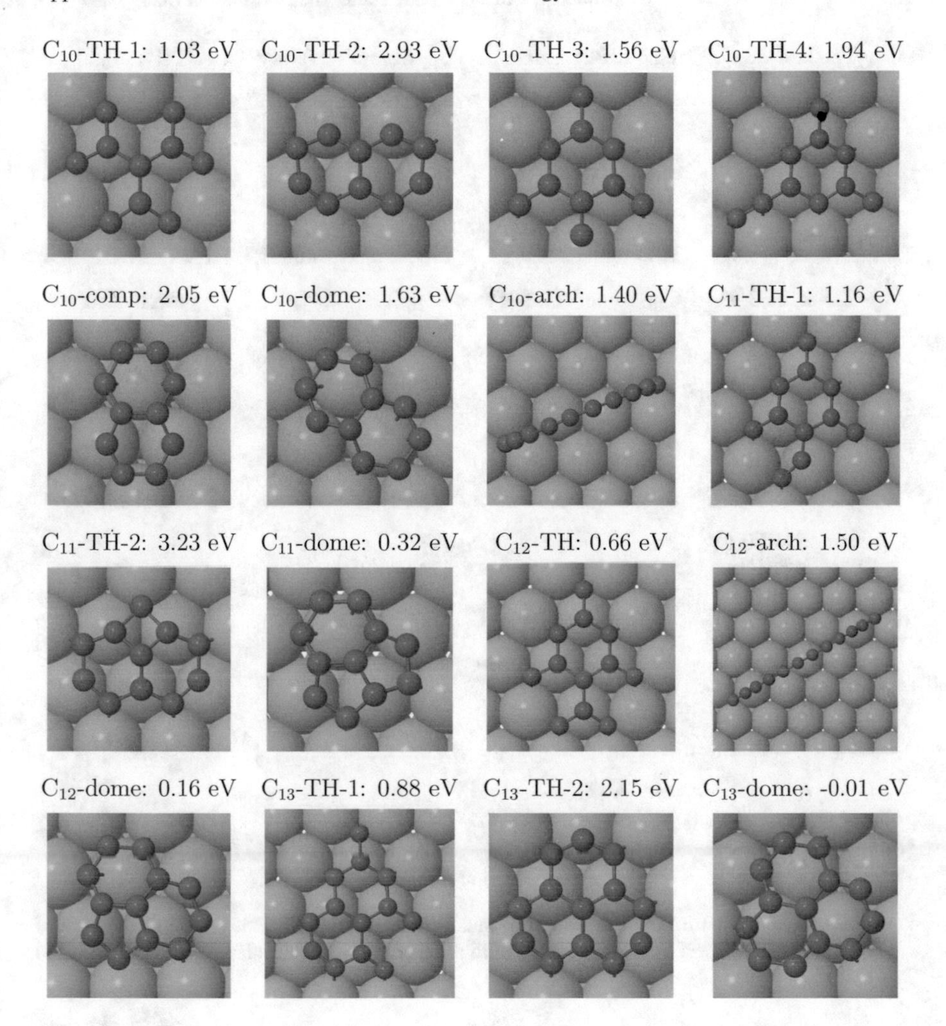
C_{10}-TH-1: 1.03 eV
C_{10}-TH-2: 2.93 eV
C_{10}-TH-3: 1.56 eV
C_{10}-TH-4: 1.94 eV
C_{10}-comp: 2.05 eV
C_{10}-dome: 1.63 eV
C_{10}-arch: 1.40 eV
C_{11}-TH-1: 1.16 eV
C_{11}-TH-2: 3.23 eV
C_{11}-dome: 0.32 eV
C_{12}-TH: 0.66 eV
C_{12}-arch: 1.50 eV
C_{12}-dome: 0.16 eV
C_{13}-TH-1: 0.88 eV
C_{13}-TH-2: 2.15 eV
C_{13}-dome: -0.01 eV

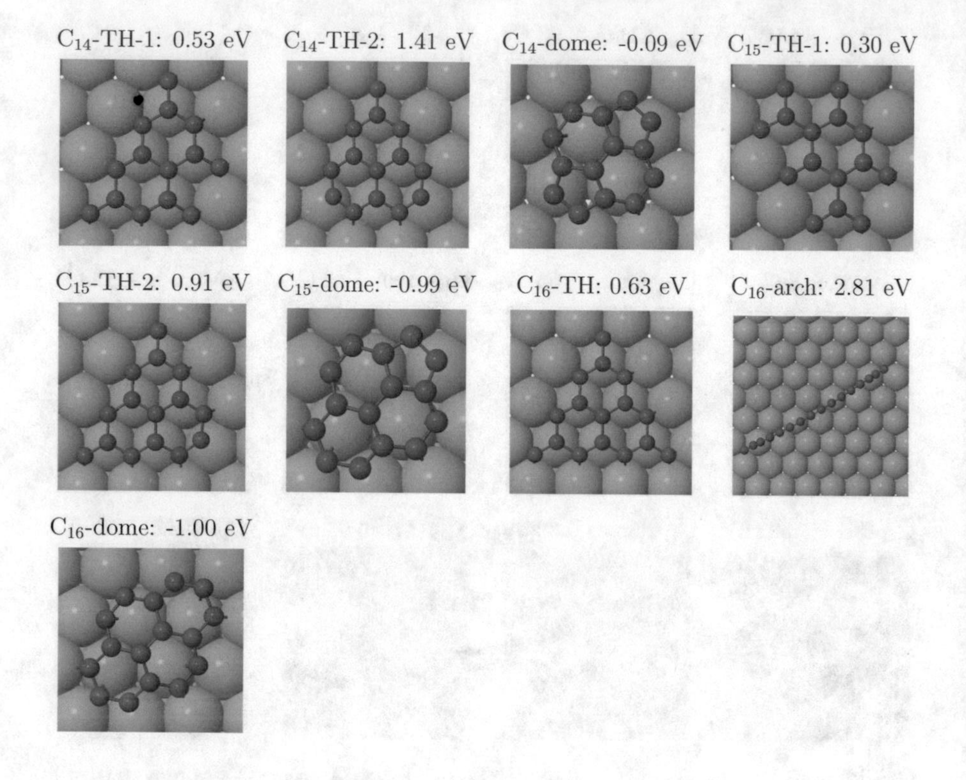

Reference

1. S.F. Boys, F. Bernardi, The calculation of small molecular interactions by the differences of separate total energies. Some procedures with reduced errors. Molecular Physics **19**, 553–566 (1970)

Printed in the United States
By Bookmasters